Population and the Environment in China

Population and the Environment in China

Qu Geping & Li Jinchang

translated by
Jiang Baozhong & Gu Ran

•

English-language edition edited by
Robert B. Boardman

Lynne Rienner Publishers • Boulder

Paul Chapman Publishing Ltd. • London

Published in the United States of America in 1994 by
Lynne Rienner Publishers, Inc.
1800 30th Street, Boulder, Colorado 80301

Distributed exclusively in the United Kingdom and Continental Europe by
Paul Chapman Publishing Ltd.
144 Liverpool Road, London N1 1LA
Tel. 071-609-5315 Fax 071-700-1057

Library of Congress Cataloging-in-Publication Data

Qu Geping
[Chung-kuo jen k'ou yü huan ching. English]
Population and the environment in China / Qu Geping and Li Jinchang.
p. cm.
Includes bibliographical references and index.
ISBN 1-55587-435-5
1. China—Population. 2. China—Population policy.
3. Environmental policy—China. I. Li, Chin-ch'ang, 1934–
II. Title.
HB3654.A3C49813 1994
304.6'0951—dc20 93-40821
CIP

British Cataloguing in Publication Data

A Catalogue record for this book
is available from the British Library.
ISBN 185396 257 0

Printed and bound in the United States of America

The paper used in this publication meets the requirements of the American National Standard for Permanence of Paper for Printed Library Materials Z39.48-1984.

Contents

Illustrations

Preface

The People's Republic of China is large in terms of both its population and natural resource base. Its more than 1.1 billion people have caused tremendous environmental pressures, resulting in pollution and irreversible damage to various ecosystems. Population and environmental problems are principal constraints to the sustained and integrated socioeconomic development of China. The government has turned to family planning and environmental protection as two fundamental national policies. In spite of some progress, multiple problems still exist; China's predicament is severe. As with many developing nations, China's growth potential depends heavily on how these dilemmas are handled.

Years of experience in environmental protection have shown that population and environmental problems are heavily intertwined. They may affect and constrain as well as promote and stimulate one another. Appropriate policies need to coordinate population growth trends with environmental affairs. Since the beginning of the 1980s, there has been an increasing number of discussions, papers, and lectures on the subject. Approached by the World Resources Institute (WRI), we wrote *A Research Guideline on Problems of China's Population and Environment*; that publication received praise from WRI and friends at home and abroad. After amendments to incorporate more information, we wrote *Population and the Environment in China*. We have attempted to include the most recent developments in population and environmental issues from both within and outside China.

Nevertheless, the subject is so broad that it was difficult to cover every aspect. There may consequently be mistakes and inappropriate points. Due to insufficient quantitative data, many conclusions remain largely hypothetical. The book is simply a preliminary attempt to expound upon the problems associated with population growth and environmental degradation. It is our hope that this book serves as a "brick," the cast of which will attract "jade." The sustainable development—in fact, the very existence—of the whole Chinese nation depends upon resolving its population and environmental problems. Our aim is to encourage people to study and help resolve China's many dilemmas. We sincerely hope that readers will give us their frank criticism.

We extend our great appreciation to all those friends who have given us encouragement and assistance in publishing this book. Our thanks are also extended to the young friends who have been collecting, selecting, and editing materials for us.

Qu Geping
Li Jinchang

Editor's Note

Dr. Qu and Professor Li's work is the result of collaborative research, writing, translation, and editing. This breadth is at once an advantage and a disadvantage. Like a message distorted in passing from mouth to mouth, the text may stray at times from the authors' original intent. Painstaking efforts have been made to minimize any such modification.

In 1990, the China Environmental Science Press published the first Chinese version of the text. Through 1992, the book underwent a sentence by sentence translation from Chinese to English. Many measurement units were converted to the metric system at that time. In 1993, I edited the translated edition for word choice, sentence structure, and general paragraph development. I additionally simplified chapter subheadings and omitted artwork deemed superfluous.

Upon completion of this third edition of the text, I conferred with Professor Li Jinchang and three translators from the International Affairs Department of the Policy Research Center for the Environment and the Economy of the National Environmental Protection Agency. Over the course of a week, we read the edited text for quantitative accuracy and reviewed all other major changes. A copyeditor further reviewed the text and submitted it to the authors for final approval. In spite of the labyrinth of checks and balances, errors in translation may still surface.

The book enhances a growing list of publications on China's population policies and environmental affairs. As administrators and advisers of China's National Environmental Protection Agency, the authors introduce a unique but commonly overlooked perspective to current dialogue on population control and environmental management in China—that of the insider. In addition, fresh quantitative data complement already published data on trends in population control and environmental compliance. In this regard, the authors' personal research is the presumed reference unless otherwise indicated. Analysis and direct commentary on implementation and enforcement of policy measures is at times lacking. For this reason, however, the book steers clear from an equally unprofessional Western tendency to dramatize.

Special thanks to the Asian/Pacific Studies Institute at Duke University and Lynne Rienner Publishers for facilitating travel expenses.

Robert B. Boardman

Population and the Environment in China

1

Introduction

China has long been a nation with a large populace. By the time the People's Republic of China (PRC) was founded, in 1949, its population totaled 540 million. Tremendous population increases in the 1950s and 1960s caused serious population problems in the 1970s. By July 1990, China's population (including Taiwan and Hong Kong) reached 1.16 billion and accounted for 22% of the world's population. In addition to the rapid increase in population, poor population characteristics and uneven distribution affect the country's development. Few countries in the world cope with such high population pressures as China's. Social, economic, and environmental problems are an inevitable consequence of such pressures.

Rapid population increases have reduced average agricultural land per capita from 0.187 ha in the early 1950s to 0.100 ha in 1991. Each hectare of agricultural land currently supports 1.78 persons, higher than average world standards. As average agricultural land per capita falls, arable land is cultivated more intensively and treated with additional chemicals. Both practices result in pollution and deteriorating land fertility. For example, in South China's hilly terrain, the organic content of soils has fallen from 6% at initial cultivation to 2% at present. Rapid urban growth has devoured large amounts of once-arable land. Even though agricultural production has increased steadily since 1949, progress has been undercut by rapid population increases. Slow increases in average food consumption in turn hinder improvements in general livelihood.

An increasing population intensifies demand for timber and firewood, thereby increasing pressures on forest resources. At present, forest coverage stands at 13% of the land area. Total standing volume is 9.12 billion m^3, or less than 9.05 m^3 per capita. Although these figures have improved over the past decade due to nationwide tree-planting programs, abusive cutting is still widespread and hampers forestation efforts. A burgeoning population's increasing demand for food also causes land conversion and damage to China's unique forest ecosystems.

China's water resources confront similar problems. Aggregate water resources total 2.800 trillion m^3, ranking sixth in the world. The average amount of water resources per capita, however, stands at 2,700 m^3—much lower than the world average. As industrial and residential water demands steadily rise, the frequency of water shortages also increases. For example, Beijing enjoyed an adequate water supply through the 1960s, but the situation changed such that by the 1970s water crises began to occur more regularly. Water shortages are now one of the most serious constraints to development in Beijing. In addition, increasing water resource pollution has caused higher demands on residual clean supplies.

Exploitation of energy resources and the resulting environmental problems are distinctly related to overpopulation. Population growth increases demand for energy, which leads to increased pressures on the environment. If annual coal consumption persists at one ton per capita, as in recent years, approximately 16 million additional tons of coal will be consumed each year due to annual births. This will considerably worsen air pollution. Rapid population growth both inhibits higher energy utilization efficiency and extends the impact of energy pollution on the environment. Gas and other convenient residential energies cannot simply be popularized in most Chinese cities because of rapid urbanization. To the contrary, millions of individual households depend on coal for heating. This has caused serious low atmospheric pollution. In rural areas, increases in commercial energy cannot match population growth. As a result, huge amounts of firewood and straw are burned for heating, causing soil degradation in rural ecosystems.

Overpopulation also degrades grasslands, mineral resources, and general living standards. In short, China's large population has become a heavy burden both to agricultural resources and to the overexploited mineral, forest, grassland, and oceanic resources. High consumer and industrial product demands contribute to the damage and degradation. At the same time, poor population characteristics counteract policy efforts to adjust population levels and environmental quality. Although overpopulation is not the only cause for China's environmental problems, it clearly plays a substantial role.

The Chinese government has carried out active and unremitting population policies due to the vitally important role that population plays in the country's socioeconomic development. China's population policies have affected family planning, population migration, marriage, and employment. Policies have long concentrated on controlling aggregate numbers. The following measures have been implemented to control and improve population figures:

1. Raise public awareness with a wide range of publicity programs that explain the importance and necessity of population control.
2. Regulate family planning so that couples are encouraged to limit family size, for example, to only one child. Government incentives and a family planning responsibility system will aid in implementation.

3. Reduce incentives for childbirth by alleviating concern for old-age care with a comprehensive social security system, guaranteeing women's health with a medical insurance system, and providing adequate contraceptive devices.
4. Promote sustainable birthrates and nursing to improve overall population qualities by prohibiting marriage between close relatives and actively promoting children's education.

Progress has been made since China implemented family planning 20 years ago. The natural population growth rate dropped from 2.3% at the beginning of the 1970s to 1.4% in 1990. This is lower than other developing nations and the world average. If the natural growth rate of the 1970s had persisted, 200 million additional people might have been born. Savings in yuan amount to RMB 2,000 billion. Additionally, population pressures on the environment were eased by an untold amount.

Serious population problems, nevertheless, still confront China. The annual 16 million newborns almost equals the population of a medium-sized country. Approximately one-fourth of China's domestic income increase has to be spent on the expanding population. This has brought about tremendous pressures on socioeconomic development as well as on environmental protection.

As with its population policies, China's environmental policies began to develop in the early 1970s, at which time many unique ecosystems had already been badly damaged. In 1973, the government began to develop environmental protection, step by step, from the "Three Wastes Campaign" to a comprehensive environmental management system with distinct Chinese characteristics. As one of China's fundamental state policies, environmental protection is carried out in accordance with urban and rural development. Specific policies are as follows:

1. Integrate prevention with treatment so that environmental protection and pollution control preempt economic development; adopt environmental impact appraisals.
2. Establish a responsibility system to clarify duties of prevention and treatment for primary abusers; renovate technology, levy taxes for cleanup, and impose deadlines for change.
3. Develop laws and regulations that strengthen responsible environmental management; establish a national management network to allocate tasks of awareness, education, and research and development at all levels of government.

China's economy developed so rapidly in the 1980s that its gross national product (GNP) doubled. In spite of signs of progress, environmental affairs have remained relatively unchanged. Some quality indicators of industrial emissions have in fact fallen. In addition, more than 2,000 organic farming experimental projects and more than 600 natural protection zones were set up throughout the nation.

China's population and environmental problems are both rooted in history; they have also become increasingly serious. The problems do not lend themselves to quick solutions—on the contrary, they mandate unremitting effort. Solutions are interwoven with socioeconomic development. China's primary objective is to continue its socialist modernization while improving population and environmental affairs. Economic development may help or hinder this task. The following strategies will help the process:

1. *Prioritize population and environmental issues with economic development:* First, appropriate economic development strategies need to direct China's floating population and promote protection of its fragile ecosystems. In the past 10 years of economic reform, a large number of laborers have been separated from farming. The government has adopted policies to employ these people within the same geographic regions—one slogan reads, "Away from land, but not from your hometown." Policies have encouraged farmers to set up village enterprises, to stimulate rural production facilities, and to avoid mass migration to urban areas. Responsibility systems for mountains, forests, and other unique ecosystems have been implemented to encourage organic farming.

Second, sustained economic development needs to be achieved with incentives as well as regulation. For example, families who have acted in accordance with family-planning initiatives should receive tax incentives or other government compensation. If land plots were allocated in accordance with the size of a family, households with more children would receive higher benefits. This sort of development policy would clearly be at odds with any population control or environmental protection plan. Moreover, enterprises might be encouraged to protect the environment if the government takes steps to "internalize their externalities," that is, businesses that pollute need to pay the price for cleanup.

2. *Develop population policies with distinct Chinese characteristics:* First, an unremitting commitment to strict population control is the country's most important criterion for development, even as the population reaches zero and negative growth rates. The natural population growth rate should decrease to 1.25% in the following decade.

Second, awareness of current population and environmental crises should be built into academic curricula nationwide. History has taught us the high costs associated with population explosions and environmental degradation.

Third, intellectual and labor resources should be exploited to adapt, regulate, and control the natural environment.

3. *Establish environmental policies that focus on population-induced problems:* First, China must insure the environmental rights and benefits of present and future populations by implementing both preventative and curative measures.

Second, public participation in environmental management must be encouraged further. At present, environmental management is mainly carried out in the name of government with little public participation.

Third, the country's vast human resource base must be used to its full potential to protect and improve the environment. The "forest belt" planted across China's northern frontier exemplifies positive human impacts on the environment. More than 9.1 million ha of planted trees have helped increase biomass production in areas struck by desertification. Large-scale grassland reclamation, urban greening, irrigation projects, and recycling programs are other ways to increase participation. An abundant human resource base is China's current competitive advantage. Effectively utilizing this resource may alleviate population pressures and pave the way to sustained economic development.

2

Population and Environmental Crises

Population control and environmental protection are two sides of the same coin. Each issue must be examined in the context of the other. They both are rooted in history, and they both have become increasingly interdisciplinary in nature. Population and environmental problems commonly surface after society is assured of survival and development. The earlier the problems are diagnosed and treated, however, the better.

From an academic viewpoint, problems associated with population growth and environmental degradation have traditionally been within the domain of demography. As theories of ecology have developed, however, these problems have stretched beyond this field. Research now covers larger areas, spans longer time horizons, and incorporates more variables. Both macro- and microscopic approaches, however, are invaluable.

Thomas Robert Malthus was a pioneer in the field of demography. He made the first comprehensive study of relationships between population growth and environmental degradation. In *Demographic Principles*, he specifically related population growth trends to land productivity. After an in-depth analysis of the dynamic relationship between the two factors, he concluded that "the growth power of population is indefinitely larger than the land's productivity." Malthus has advanced a hypothesis that whereas populations may increase geometrically, mankind's means of subsistence may increase only arithmetically. In addition, he has expounded upon the following three propositions: (1) means of subsistence will restrict population growth, (2) population will grow as the means of subsistence grows, and (3) poverty and general malice will restrain potential population growth so that actual growth balances with the means of subsistence.

The Malthusian approach to population issues is rather simple. In spite of severe criticism, its many implications have had a tremendous impact on the study of demography. Malthus has forever impacted research on population growth, environmental quality, and general productivity. His analytical methodology and interdisciplinary approach have been replicated and critiqued by researchers worldwide.

Since the 1950s and 1960s, population and environment issues have become a global concern as population crises have generated natural resource and other environmental dilemmas. The issues have become focal points for competing scientific theories on population trends. Two such theories, or schools of thought, might be classified as "the pessimists" and "the optimists." The two theories have deep social implications.

Representatives of the pessimistic theory on population include U.S. scholar William Foster, demography experts W. S. Thompson and J. H. Hertsler, ecologist Laol R. Ellish, and scholars Jay W. Frost and D. H. Maydos. Their pessimistic view holds the following:

- Many regions of the modern world, especially developing nations, are plagued by excessively large populations and their impacts on natural resources and general productivity.
- Rapid consumption and depletion of various natural resources potentially leads to resource exhaustion.
- Population explosions bring humankind to the brink of destruction.
- Excessive populations and increased resource consumption will cause the world's demise; economic growth is limited, not limitless.

Representatives of the optimistic school of thought on population include U.S. scholars Herman and Cain; and economists Julian, Lincohn, and Sieman. The main views of this school are as follows:

- Mankind's use of nature for development is limitless.
- Limitless energy and resource potential make economic growth forever sustainable.
- Science and technology will overcome most human and economic obstacles.
- Free markets are an efficient mechanism that adjusts utilization of natural resources so that crises will be averted and solved automatically.

In general, pessimists regard sharp population increases as the main source of human crises, whereas optimists deny the direct relationship between population and human crises, or simply deny the existence of crises altogether. As presented here, the two viewpoints are overly biased and extreme. Crises of any nature call for cautious, calm, and unbiased scientific analysis. Heated arguments that stem from contrasting viewpoints have admittedly, however, aroused constructive concern and in-depth deliberation over the issue. The debate has helped to yield valuable understanding and creative initiatives.

The interface of demography, environmental sciences, and economics has become a rich source of research. Researchers and the general public have become increasingly aware of the problems and increasingly optimistic about the solutions. Macro- and microanalyses have stretched beyond the basic Malthusian framework.

China has long been distressed with population and environmental problems. Continuous deliberations concerning population and environmental issues stretch back to the Spring and Autumn periods in Chinese history. Moreover, Mencius commented on the aesthetic and environmental effects of clearcutting practices along the Yellow River in the fourth century B.C. The debates have continued through the Ming and Qing dynasties to the present.

In the mid-nineteenth century, China had a population in excess of 400 million, the maximum limitation for a feudal small-scale farming economy. According to Karl Marx, "the continuous increase of excessive population in China had long before made the social conditions heavy yoke for the majority of the nation." Against this grim commentary, almost every eminent thinker of modern times in China since Gong Zizhen has proposed solutions to solve the population problem. Suggestions are commonly made to ban early marriage and encourage migration to Northwest China.

The 1911 revolution popularized Malthus in China. Open deliberations molded two contrasting viewpoints. One held that China's vast territory and expansive resource base were more than adequate to support China's population. The other contrarily viewed China's population as excessive and the country's resources inadequate. With the entire nation on the verge of disintegration, thorough resolutions were impractical.

The social context for constructive dialogue and objective analysis was slow to develop. After the founding of the People's Republic of China in 1949, the Soviet Union heavily influenced affairs in China. Malthus and ideas on geographic determinism were severely criticized and suppressed. There were few deliberations in this regard until the early 1970s.

In the 1950s, the famous Chinese economist, Ma Yinchu, published *New Demography*. With a set of assumptions that run contrary to Malthusian thought, the book proposed solutions to the population problem. It recommended that population control be implemented to solve shortages in the food supply and industrial raw materials. It furthermore promoted scientific research and technological development. Ma suggested the following objectives: improve the population quality, control population growth, practice late marriages and birth control, limit the number of newborns per family, and incorporate population control measures into the national plan. Ma clearly recognized the contradiction between population and productivity in China, but he was severely condemned for upholding his views. As a result, the foresight and sagacity of Ma and others failed to spur social action and scientific inquiry into population control.

Since the 1970s, social demand for population control and environmental protection has popularized demography and environmental sciences in China. As foreign and domestic theories infiltrate China, environmental impact assessments and population forecasts have become increasingly commonplace. The public warmly received a research report by the Chinese Academy of Sciences, *Survival and Development*, which presented a comprehensive list of recent domestic concerns involving population and the environment. Additionally, the State Planning Council's Committee of Comprehensive Investment in Natural Resources recently conducted a survey on the productive and carrying capacity of China's land resources.

After years of research and management of natural resources, we feel strongly that population growth is the primary constraint on overcoming China's environmental problems. An excessive population leaves China with little room to maneuver and solve problems. If people are compelled to farm mountainous slopes, harvest trees unsustainably, mine ores blindly, and pollute water resources endlessly, the potential for further development will be bleak.

Our contemplation and deliberation rest on the following three points: (1) evidence of environmental impacts, (2) mechanics of the population-environment relationship, and (3) suitable measures to address the problems. First, the population problem is expressed by the pressures it exerts on the environment. We should therefore first examine the interaction between China's population and the environment. Populations do not pressure the environment at all times and at all places. The pressure is felt when population grows past a certain level. Before population and environmental crises attracted the world spotlight in the 1970s, sporadic environmental impacts occurred throughout the 1950s and 1960s. More recently, impacts have been detected on a global scale. Rapid population growth and industrialization, compounded by resource depletion, acid rain, destruction of the ozone layer, and climate change, have caused worldwide concern. Population pressures on the environment are now commonplace and widespread. With a population of more than 1.1 billion, China is especially vulnerable.

Second, population pressures on the environment are frequently indirect. Humans unquestionably develop and utilize resources, thereby altering the environment. The degree of change is a function of science and technology, organizational modes of production, and timing. It is therefore almost impossible to attribute the brunt of the blame to any single factor. Appropriate preventative and curative measures to solve population and environmental problems depend on research that examines the multiple interactive processes. A clear understanding of the mechanisms of the population-environment relationship is of utmost importance.

Third, understanding the problem is a prerequisite to solving it. Before the 1970s, a blind view prevailed in China that gradual increases in population would improve a socialist economy. In the 1970s, these ideas changed radically. Active population control and environmental protection consequently took root. In the

1980s, family planning and environmental protection were deemed "basic state policies."

Population control and environmental protection efforts in China will require unremitting efforts for ages to come. Appropriate resource utilization strategies and overall guidelines for socioeconomic development will require effort and patience on the part of all involved parties. In this context, further analysis and evaluation of China's present and future policies constitute our third concern.

3

Population and Environmental Issues Throughout China's History

Human development has largely been a function of humankind's relationship to nature. Throughout the years, environmental impacts have depended on fluctuations in population growth, advances in science and technology, and increased public awareness.[1] A brief survey of the relationship between population growth and environmental affairs in Chinese history lays the foundation for present and future research and progress.

Throughout human development, understanding of nature has spiraled from fear to disregard. The process has generally followed changes in human environmental values. The degree of knowledge of nature's many systems determines changes to the environment. Before the Industrial Revolution, low productivity caused minimal effects on the environment. Population increased only gradually throughout 4,000 years of Chinese history and then increased rapidly in the nineteenth century (Figure 3.1 and Table 3.1).

The history of population and environmental affairs in China can be divided into five stages: the pre-Qin "Golden Era of the Environment," the Qin to the West Han "Signs of Degradation," the East Han to the Sui "Signs of Recovery," the Tang to the Yuan "Secondary Degradation," and the Ming to 1949 "Escalated Degradation."

Pre-Qin Period: "Golden Era of the Environment" (2205 B.C. to 221 B.C.)

The pre-Qin refers to the period from the Xia dynasty, the first slavery state founded in 2205 B.C., to the start of the Qin dynasty's united feudal empire in 221 B.C. The period spans approximately 2,000 years. Throughout this period, China's population grew slowly from 10 to 20 million people. It was a sparse

Figure 3.1 Chinese Population, 20–2000 A.D.

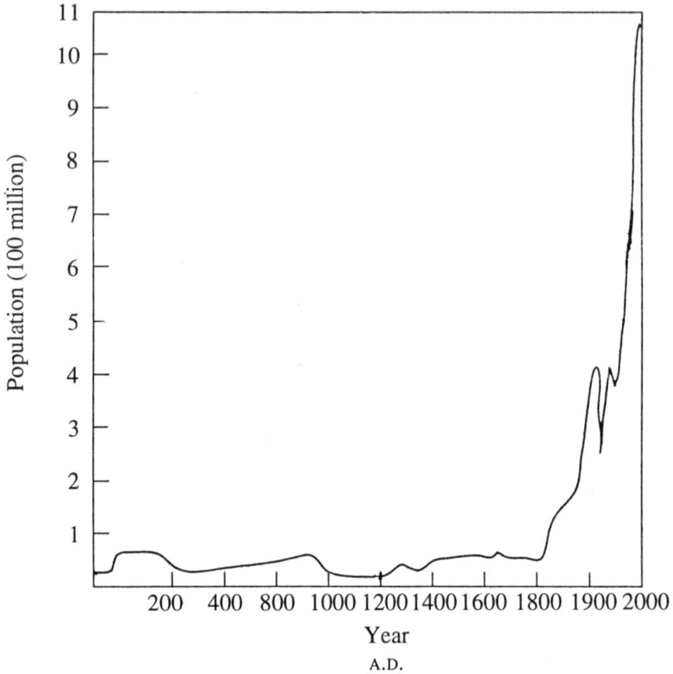

population relative to China's vast territory. Recent excavation of relics from the pre-Qin indicate that the population was evenly distributed throughout the land. There are traces of human activity along almost every river and lake and on every grassland. Three people—less than one family—existed per square kilometer. Few families lived in solitude. Centralized habitation may have been a means to lessen the effects of natural disasters and maintain daily productivity. Most areas of the country were uninhabited at the time. In later eras, the Yellow River Valley became the most centralized and densely populated area.

A population of 10 million inflicted minimal damage to China's environment. Poor technology, low productivity, and backward modes of production minimized the effects of humans on nature. Myths from the period reveal the predominance of nature and the fortitude of humans. The "Mountain and Seas Scripture" tells a story of how "at the time of Yao, ten suns rise all at once, burn the crops, ruin the plant and leave nothing to the people for food . . . then Yao sent Yi to shoot nine of the ten suns . . . and people were happy thereafter." The tale reveals that people at the time were defenseless in the face of drought and other natural disasters. Only their imaginations conquered and controlled nature. This and other myths express the will of the people to conquer nature.

A small population and low productivity, then, kept China's environment relatively undisturbed. Even the Yellow River Valley, with its increasingly dense population, remained highly vegetated and extremely productive through-

Table 3.1 Changes in Population and Environment in Chinese History, 2205 B.C. to 1949 A.D.

Stages	Periods	Years	Representative Dynasties	People (10,000)	Status of Environment
I	Pre-Qin	2205–2198 B.C.	Xia and Yu	1,355	"Golden Era"
		1115–1079 B.C.	West Zhou (King Chen)	1,371	
		684 B.C.	East Zhou (13th Year, King Zhuan)	1,184	
II	Qin to West Han	221 B.C.	Qin (26th Year of Emperor Shihuang)	2,000	Signs of Degradation
		2 A.D.	West Han (2nd Year)	5,959	
III	East Han to Sui	57	East Han (2nd Year Jianwu Zhongyuan)	4,500	Signs of Recovery
		220–280	Three States	4,000	
		606	Sui (2nd Year of Emperor Yang)	4,601	
IV	Tang to Yuan	755	Tang (14th Year of Tianbao)	5,291	Signs of Secondary Degradation
		1195–1223	South Song (1st Year of Quingyuan)	7,881	
		1290	Yuan (27th Year of Zhiyuan)	5,883	
V	Ming and Qing to before liberation	1403	Ming (1st Year of Yongle)	6,659	Escalated Degradation
		1651	Qing (8th Year of Shunzhi)	5,300	
		1684	Qing (23rd Year of Kangxi)	10,170	
		1762	Qing (27th Year of Qianlong)	20,047	
		1790	Qing (55th Year of Qianlong)	30,148	
		1834	Qing (14th Year of Daoguang)	40,100	
		1919	Republic of China (8th Year)	50,600	
		1947	Republic of China (36th Year)	54,887	

out the Xia, Shang, and Zhou dynasties. South of the Yangtze River, land was sparsely populated and largely pristine. Needless to say, other factors help explain the healthy maintenance of China's pre-Qin environment.

The West Zhou (1115–1079 B.C.) was the first dynasty to take measures to protect China's environment. The government focused on the protection of wildlife habitat. According to "Rite of Zhou: Regional Officer," the responsibilities of local governors included how locals treated their surroundings and the protection of rivers, mountains, forests, birds, and other animals.

The Xia, Shang, and Zhou dynasties witnessed regular migrations. According to historical records, the Xia people migrated ten times, the Zhou seven times, and the Yin (of the Shang dynasty) migrated sixteen times. The migrations helped ease pressures from excessive centralization.

Far-sighted politicians during the pre-Qin, such as Guan Zhong of the Qi State in the Spring and Autumn periods, advanced causes for the environment. He approved of increases in population with conditions. He held that rulers must control and arrange populations in accordance with two key ratios; otherwise large populations would induce natural disasters and national instability. One of the ratios measures land and population. Guan Zhong reasoned that a lot of land allocated to few inhabitants would cause inefficiencies in land use. Relatively small amounts of land allocated to a greater number of people, on the other hand, would cause inefficient farming and food shortages. "A good ruler," he concluded, "knows well how many people he reigns over; only sufficient arable land will make his people content."

The second ratio measures different components of the population. Guan Zhong specifically noted the proportion of urban versus rural, and military versus civilian population groups. An excessively large urban population, he reasoned, would degrade the rural population's living standards. In such a scenario, "there would not be enough farmland to support the population." If means of subsistence do not meet the needs of the population, this might lead to nationwide poverty and hunger. Moreover, a disproportionately large military would "leave the vast land uncultivated and also bring the nation to the brink of poverty." Guan Zhong's foresight retains its value even today.

Due to population levels, productivity, people's awareness, and government measures, China's pre-Qin environment was left intact. The middle reaches of the Yellow River were characterized by vast plains, a moderate climate, and dense vegetation. Forest coverage of the Loess Plateau exceeded 50%; the most dense forests were in the central Shaanxi plain. The lower reaches of the Yellow River had a low, smooth terrain with few hills and many lakes. The climate was humid and the vegetation abundant.

Although agriculture in China has a history of 5,000 to 6,000 years, technological innovations were limited during the pre-Qin. Human activities therefore had little impact on the environment. Nature, on the other hand, strongly influenced population levels. Fluctuations indicate that the environment had more of an effect on China's population than the population had on the environment.

Qin to West Han Dynasties: "Signs of Degradation" (221 B.C. to 57 A.D.)

"Six states conquered, the whole world is united." In 221 B.C., the first emperor of the Qin united six states and established the first feudal centralized nation in Chinese history. Fighting that had lasted for centuries among the warring states finally ended. The unification of the Qin, however, was beneficial neither to population control nor to the dynasty's environmental affairs.

Unification by military force killed more than a million people and lay waste to 13 cities. Deaths related to hunger and displacement outnumbered those related to battle. After unification, large-scale construction detracted from other productive activities, leaving people with little respite. About 400,000 people helped build the Great Wall. Another 500,000 guarded mountains and suppressed riots. And 700,000 built the E'fang Palace and Qin Shihuang's tomb at Lishan Mountain (site of the Terracotta Warriors). Diggers became sacrificial objects themselves. As a result, there were an insufficient number of men for farming and women for weaving. Although unification of the Qin was a great contribution to history, its rulers extorted taxes and forced military service on the people. The population of the Qin fell to below 20 million.

Construction efforts during this period caused large-scale environmental degradation to forests and other natural resources. The poet Du Mu of the Tang dynasty approved of unification and the glory of E'fang, but lamented deforestation: "Six states conquered, the world was united . . . with Shushan Mountain bare, E'fang Palace rose to glory." The short reign of the Qin did not allow enough time for the population to recover. The population, therefore, had little direct effect on the environment. Environmental degradation was largely a function of construction and deforestation.

The West Han dynasty witnessed increases in population and added pressures on the environment. This dynasty was founded by Liu Bang Gaozu of the Han. Feudalism achieved unprecedented prosperity under the reigns of emperors Hui, Lu, Wen, and Jing. Records from the Han dynasty relate how "granaries overflowed and every household increased savings." Social stability and economic prosperity compounded with Emperor Hui's birth incentives increased the population to 59 million, a record level that remained unsurpassed for centuries. A fine levied on women over age 30 who failed to produce children exemplifies the dynasty's birth incentives. A prosperous population symbolized national prosperity. Sharp rises in population, nevertheless, caused China's first widespread environmental degradation.

This early degradation is commonly attributed to two principal factors. First, population growth increased demand for food and exacerbated pressures on the land. To feed an increasingly large population, the existing arable land was no longer sufficient. Cultivating wasteland hence became the primary solution to the dynasty's food shortages. Approximately 8.27 million ha of land was reclaimed for farming.[2] Although productive during the initial crop rotations, land rapidly lost productivity, as happened in the Buhe Desert.

Formerly productive during the West Han, this area produced enormous amounts of grain and then lost its fertility because of poor management. In the Song dynasty, the area had become, "a sandy place that people couldn't cross but by camel."[3]

Other areas also underwent large-scale cultivation. During Emperor Wu's reign of the Han, 700,000 people migrated to cultivate the Huangtu (Loess) Plateau. The once-nomadic plateau has since changed into agricultural land.[4] In the vast areas south of the Yangtze River, lands also underwent cultivation. Farming in the middle reaches of the Yellow River destroyed forests and grasslands, resulting in serious soil erosion and changes in the river course. River beds rose in the lower reaches of the Yellow River; dense forests disappeared in the central Shaanxi plain; and grasslands decreased nationwide.

Second, papermaking and iron-smelting technologies during the West Han greatly increased the human impact on the environment. Unearthed relics in Gong County, Henan Province, revealed 18 iron-smelting pots within a range of 1,500 m^2. At another site in Nanyang, 17 pots were uncovered in a 3,000-m^2 area. Such large-scale iron-smelting was previously unknown. Cattle farming also became increasingly popular during the West Han, further contributing to human impact on the land.

East Han to Sui Dynasties: "Signs of Recovery" (57 A.D. to 755 A.D.)

As populations decreased, environmental affairs in China began to recover from the East Han to the Sui dynasties. This was one of China's more turbulent periods. Battles and other disasters caused populations to fluctuate. Gauging these fluctuations against changes in environmental affairs might further shed light on the relationship between population levels and environmental affairs.

The East Han dynasty was established on the ruins of upheaval. Only 2 of every 10 people survived the period from when Wang Mang usurped the throne to the reign of Emperor Wu Guang.[5] By the second year of Emperor Wu Guang's reign, the population had fallen to 21 million people. In the subsequent 50 years, the economy recovered, however, and population increased. By 98 A.D., the population again exceeded 50 million under the reign of Emperor He.

Peace and stability were interrupted again, however, from 107 to 135 A.D. under emperors An and Shun. The East Han became less authoritative as land was annexed and differences between rich and poor grew. Each of these factors contributed to the dynasty's demise. Peasants suffered and finally revolted under exploitation by despots and landlords. From the reign of Emperor An to the end of the East Han, more than 100 peasant uprisings took place. Social chaos worsened in 150 A.D. after the reign of Emperor Heng. The Huangjin and Dongzhuo uprisings contributed to increasing death rates, as separatist warlord regimes scrambled for supremacy.

In the Dongzhuo trurmoil, Tai Zu (Cao Cao), ruler of the tripartite Wei State, blocked the Si River and drowned tens of thousands of peasants. According to "Records of the Three States," the army attacked and robbed people who would otherwise have died of hunger. Locusts and droughts further contributed to the death toll. After returning from Shaanxi to Henan, the emperor himself had little to eat. Wang Jie's poem describes how "nothing can be seen but bones over the plain." Cao Zhi further describes how "the central plain is desolate with few signs of human habitation." Even Tai Zu wrote in a poem that "bodies are seen here and there, roosters crow nowhere; one is alive and ninety-nine dead, what a sad scene it has made." These are but a few of the numerous literary descriptions of the period. The population at the time totaled approximately 10 million people. Nearly three-fourths of the population under the East Han had died off in turmoil.

In 265 A.D., the Wei, Shu, and Wu united under the West Jin dynasty. Population growth picked up slowly, however, due to continuous warfare. Only 17 years passed before civil turmoil broke out again; it was discovered that Sima's kings involved themselves in the "Eight Kings Turmoil," which lasted 16 years. Population growth rates dropped dramatically after only a short period of increase.

In the fourth century A.D., China fell once again into turmoil. North China became a battlefield for various nationalities, whereas dynasties in South China rotated power—the Song, Qi, Liang, and Chen states followed the East Jin dynasty. The Yellow River Valley throughout this period was in unprecedented chaos. Fragmented for a long period, the population fell dramatically.

Unification under the Sui dynasty in 581 A.D. again paved the way for revival and development led by agriculture. Society placed economic growth in high priority; the commitment would not last long. By the second year of Emperor Yang's reign, China's population again rose to 46 million people. With a taste of stability and prosperity, the emperor's appetite loomed large. He engaged in large scale construction and wanton military aggression. Exorbitant levies and unnecessary death and destruction were the prices paid. China's population again tumbled.

From the East Han to the Sui, China's population rose and fell for 600 years. People suffered primarily from warfare, and they lived tempered and tortuous lives. The sharp drops in population, however, served to alleviate pressures on the environment. A time-worn analogy of a clam, a crane, and a fisherman may best describe this trend: "If cranes attack clams, fishermen benefit." In other words, when a clam opens its shell under the sunshine, cranes will attack and feed. If the clam bites back, fishermen will harvest both. The story suggests that if two parties are at war, a third party may reap all the benefits. In this event the two parties are the Chinese at civil war and the third party is the environment.

Warfare frequently interrupted periodic unification. Population declines, general chaos, fallow farmland, and southward migrations hurt agricultural productivity. Plains in the Yellow River Valley expanded as forests were overharvested and reclaimed for farming. Although fires from military activities caused destruction, they were limited to certain areas and by no means severe.

Vegetation of the middle reaches of the Yellow River was still relatively well protected. Less severe soil erosion afflicted the lower reaches of the river. Before technological breakthroughs, then, it appears that a rather direct relationship existed between China's environmental affairs and fluctuations in the environment.

Tang to Yuan Dynasties: "Secondary Degradation" (755 A.D. to 1403 A.D.)

In the Tang dynasty, feudal society experienced prosperity and good government. Poets recorded life during the Tang in descriptive verse. Improved living conditions and government policies encouraging large families caused the population to rapidly expand. In addition, large numbers of people immigrated from other countries to China so that by 755 A.D. the population reached more than 53 million.

The country did not experience genuine unification in the Song dynasty. With the Yellow River as its southern border, the North Song coexisted with the Liao and Xixiao states. The South Song later retreated to areas south of the Huaihe River and lived in conflict with the Xixia and Jin. Finally, the South Song settled in the areas south of the Yangtze River to seek peace and contentment. For a long period, the dynasty subsisted and compromised its resources to ever-worsening conditions. China's total population reached 76 million people by 1223 at the end of the South Song.

As the Song dynasty fell, Mongolian nationalism rose. Tired of their existence in the Yinshan Mountain region, the Mongols began aggressive military advances southward. Breaking through the Great Wall and crossing the Yellow and Yangtze rivers, they unified China and established the Yuan dynasty, the first minority-ruled dynasty in Chinese history. During this period, the population stabilized at between 58 and 65 million. In spite of occasional nationality confrontations and general nationwide stability, population levels never exceeded this range. China's population distribution and general economic prosperity, however, changed significantly.

The Northwest

To strengthen the rule over the Northwest, the Tang dynasty set up garrison cities at Anbei, Beiting, and Anxi. It also established 6 cities and 7 districts north of the desert. To make the inland more easy to access, it built a road along which there were 68 outposts. These administrative measures strengthened economic and cultural relationships. Exchange between the Northwest and the coastal areas flourished, creating favorable conditions for population growth. The region became more densely populated and cultivated. The Song dynasty did not include the vast northern and western territories. The three garrison cities became inactive as the Tang government fell apart. Southern states such as Tufan and Nanzhao rose and collaborated with the Song dynasty. The population in the Northwest increased, but relatively less than that of the entire country.

After the South Song and up until the Yuan dynasty in 1271, Mongolian nationalism wreaked havoc in the North and Northwest. On horseback, the Mongols chose first to conquer these areas from which they could move south to conquer the Jin and Song minorities in the Northwest. The minorities were commonly killed or forcefully thrown into the military. Large-scale emigration to the South, declines in western agricultural production, and a decrease in the population level ensued.

The Yellow River Valley

The Yellow River Valley is considered the birthplace of China's civilization. It was the economic, political, and cultural center for every dynasty through the Tang. Chang'an (present day Xi'an) was the capital of the Tang and was home to more than 1 million people. It became home to indigenous minorities, as well as Japanese, West Asians, and even Europeans. The primacy of ancient China, Tang civilization was subsequently wrecked by natural and anthropogenic calamities, frequent wars, and general turmoil. The prosperous Yellow River Valley became desolate; water-saving facilities and farmland lay to waste; masses emigrated south or starved.

Since the end of the Tang, environmental degradation in the upper reaches of the Yellow River has caused frequent floods. Until the coexistence that ensued in the Period of the Five States, this region had acted as a battlefield for more than 50 years. Rulers of the 5 different states would breach the Yellow River to win military victories. Needless to say, their actions caused untold damage to the people and severely destroyed water projects in the North. Such socioeconomic decline induced general population decreases

The Yangtze River Valley and the South

In the northern areas of the Song, frequent peasant uprisings broke out. The Jin State invaded during continuous turmoil and scorched the earth wherever they traveled. Northerners began to follow the South Song migration from the Yangtze to the Pearl River Valley. Whereas the North experienced degradation, migration, and economic ruin, the South rapidly developed. Invasion by the northern Jin State effectively promoted development in the South. Large-scale immigration enhanced productive forces in the South and promoted cultural and technological exchange. Farmers immigrating from the North made full use of the South's favorable climate and soil fertility, developing efficient irrigation systems and continuously increasing grain production. Hence, the Yangtze Delta, Jiangxi, Sichuan, Hunan, Guangdong, Guangxi, and the Pearl River Valley became the predominant grain-producing areas. In Suzhou, rice production averaged 1,500 to 2,250 kg per ha in normal years. Production was higher in more southern areas because of multiple crop rotations and other factors.

In the meantime, the South Song made surprising advances in marine transportation. Foreign trade in the southeastern coastal provinces fueled economic development. In sum, the Song and the Yuan witnessed increasingly dense populations in the South and the Southeast and increasingly sparse populations in the North and the Northwest. Changes in population density and socioeconomic development during this period exerted a profound influence on the environment.

In spite of advances in agricultural technology, people lacked concern for environmental protection. The rise and fall of populations might have acted as a barometer to estimate environmental degradation. Any increase in population generally suggested massive land reclamation. In the heyday of the Tang dynasty, food production fell short of the needs of 53 million people. To meet the shortfall, 6.2 million ha of land were reclaimed for agriculture, severely damaging many fragile ecosystems. Part of this area includes the Huangtu Plateau. Unique vegetation types on the plateau were sacrificed for animal husbandry. Soil erosion, river expansion, and flooding were other unintended costs. The Yellow River overflowed its banks more than 50 times in 100 years of Song dynasty rule.

Ningxia Hui Autonomous Region, the eastern part of Gansu Province, and the northern part of Shaanxi had been a nomadic area in the East Han. After several hundred years, vegetation recovered and desertification ebbed. During the Tang, the Dang Xiang nationality, led by its political leader Tuoba, began to reclaim farmland in the area. After centuries of laying fallow, the land turned productive under Dang Xiang's people and populations increased. The result was a new government of West Xia in 1038 that acted in competition with the Song dynasty. The West Xia government, courageous and capable of competing with the Song for more than 100 years, necessitated large supplies of grain. Within the 300 or more years of West Xia government, large-scale farmland reclamation and a transformation from animal husbandry to farming severely destroyed the region's environment.

The overflowing of the Yellow River induced desertification in North China. The Maowusu Desert spread to the Great Wall, forcing the ruling center to move westward from Xia, Sui, Yin, You, and other states to Yinchuan City on the banks of the Yellow River. The West Xia began to decline thereafter. The rise and fall of the West Xia was reflected in changes to the region's environment. Recovery of the environment, in turn, made the region's population increase and economy boom. Reclamation and cultivation over one or two centuries, however, brought the dynasty down with increasing desertification and other natural calamities. Invasion by the powerful Mongols accelerated the downfall of the West Xia dynasty.

The Song dynasty witnessed a second population peak near the end of its reign. In 1223, the population reached 76 million. When farmland was in short supply, farmers cultivated land around lakes and mountains. As a consequence, soil erosion grew more severe. Loss of the territories in the North intensified land reclamation in the South. Reclamation during the Yuan in the following century amounted to 170,000 ha. Ecological diversity suffered in the North and South. The Yuan was the only dynasty that lacked agencies for protection of mountains, forests, rivers, and lakes.

In conclusion, sharp reductions in forested areas and lakes, increasingly severe soil erosion and desertification, and more frequent flooding by the Yellow River comprised the second period of environmental degradation in China's history. Changes in population during this period are also noteworthy. Once slightly more than 17 million in the mid-Tang, the population rose to 46 million and fell to 16 million during the Song dynasties. Causes for these fluctuations were primarily war-related. Differences over land availability played a major role. Moreover, although the people were increasingly capable of affecting nature, they were equally ignorant of nature's complex systems. An average 0.67 ha of farmland per capita was required to ensure sufficient food production. Peasant uprisings ensued whenever population and environmental affairs were thrown askew. The uprisings, in turn, caused lasting chaos and famine, population declines, and migration.

Ming Dynasty to 1949: "Escalated Degradation" (1403–1949)

The 600 years from the Ming dynasty to the Communist victory in 1949 laid the foundation for China's current population problems. Heavy population pressures throughout this period considerably impacted the environment.

The Ming began its reign by capitalizing on the fruits of peasant uprisings. By the end of the Yuan dynasty, destitute and homeless masses made any national population census nearly impossible. Zhu Yuanzhang undertook the Ming's first comprehensive population survey in 1381. By registering households within the dynasty's domain, Zhu estimated a population of 59 million. Twenty years later, in 1403, the population reached 66 million, the highest figure in Ming records. At the time, the political situation was relatively stable and the economy was at its prime. An unprecedented commodity economy developed as capitalism began to spread nationwide. "Floating" manual laborers and entrepreneurs offered their services and slowly migrated with others from rural inland areas of the country to more urban coastal areas. Literature from the period and state records describe the burgeoning population, the commodity markets, and the mass migration.

In the Ming dynasty's final years (mid-seventeenth century), population levels must have been higher than the peak 1403 figure. Zhao Wenlin and Xie Shujun estimate that the population may have approximated 100 million.[6] Chen Caizhang more conservatively estimates a peak of 85 million people. Lu Zhenyu contrarily figures that the population may well have exceeded 100 million people, based on Song dynasty and earlier Ming dynasty records. It would be fair to assume that the population well exceeded the 60-million mark.

Population rose dramatically during the Qing dynasty from 1644 to 1911. Policies in the Kangxi period encouraged birth. Moreover, the dynasty distributed land according to the number of people in a family. Further, it abolished the poll tax. Eight years after the suppression of peasant uprisings in 1643, the population totaled 53 million. Thirty years later, it exceeded 100 million; in 1762, it exceeded 200 million; and in the 1790s the population grew to more than 300 million. Several years before the Opium Wars, in 1834, the population exceeded 400 million. In spite of the turmoil associated with the Opium Wars, the 1911 revolution, and warlordism, the population continued to increase exponentially with its large base and government incentives. In 1919, China's population exceeded 500 million. Neither civil wars nor imperialist invasions had much effect on the population. Lao She, a famous Chinese playwright, wrote, "the killing by the Japanese could never match the birthrate of the Chinese people." On the eve of the Communist victory in 1947, the population totaled 548.87 million.

Backward agricultural technology and productivity exacerbated environmental impacts from such an enormous population. Even Emperor Yongzheng realized the serious problems that large populations may bring. To avoid massive unemployment and social instability, the emperor encouraged expanding China's arable land base. The foresight, although not environmentally sound, is admirable and noteworthy. Large-scale cultivation was encouraged as early as the fifteenth century. According to statistics from 1383, reclaimed land exceeded 1.8 million ha, or just over half of the nation's farmland. By 1393, reclaimed land rose to 8.5 million ha. At the time, the military joined hands with entrepreneurs in reclaiming land. Every able individual was mandated to cultivate 6.7 ha. The cultivation achieved temporary benefits, but at long-term costs to valuable ecosystems. Farmland nationwide amounted to 4.29 million ha. Vast tracts of newly reclaimed land were left to waste.

New approaches to agricultural production wreaked havoc on the environment. Dikes surrounded low-lying fields. Entire forests were harvested for cultivation. Reclamation additionally occurred on the frontier, prairies, mountains, and islands—all unique areas that buffered densely populated regions. Under Emperor Qianlong, farmers cultivated 21,300 ha and reclaimed another 33,300 ha within an 11-year period in the Northwest. Uncontrolled population growth and land reclamation exceeded sustainable development in the region. Assuming that an individual subsists on less than 0.27 ha, China's population reached a critical level in 1754. From this point on, affairs worsened throughout the Qing. Under heavy pressures, the environment deteriorated considerably in subsequent centuries.

Major signs of environmental degradation included, first and foremost, deforestation. On the Huangtu Plateau, forests near and along the Weihe River, the Hengshan Mountains, the Erduosi Plateau, and the Yinshan Mountains, as well as Kelan, Wuhai, Baode, Pianguan, and Hequ counties, were largely destroyed. Deforestation north of the Qinling Mountains exemplifies the destruction. The rapidly increasing population of peasants from central Shaanxi, Hunan, Guizhou, and Sichuan entered the forests to reclaim and cultivate land. Employing various harvesting methods, the peasant farmers would leave nutrient-rich soils infertile after several years and deforest new areas. Centuries-old forests north of the Qinling Mountains were left treeless. Forests around Beijing and the lower reaches of the Xianjiang River were also destroyed in the Ming and Qing. In the first few years of the Ming, the West Mountains of Beijing were well forested. After centuries of timber harvesting, the mountain slopes were nearly bare.

An additional sign of environmental degradation includes unprecedented soil erosion. Sand content in the sedimentary layers of the Yellow River reflects soil erosion and frequent overflows. During the Ming dynasty, the river overflowed 127 times, approximately once every 2 years. Later, during the Qing dynasty, the river continued to flow over its banks, flooding more than 180 times in a 200-year period. The Yellow River overflowed substantially less in prior periods. Flooding averaged one in 26 years during the Qin and Han dynasties and one in 10 years during the Three Kingdoms Period and the Period of the Five States.

Cultivated areas and cities of the Han and Tang dynasties in the North and Northwest were literally submerged by sands during the Ming and Qing reigns. Both the Dunhuang Grotto and the Silk Road are surrounded by a seemingly endless sea of sand. The Badan Jilin, the Ulanbuhe, and the Maowusu deserts continuously expanded and destroyed vegetation. Desertified areas of Kergin in Northeast China were more or less created by land reclamation during the Qing. Lakes dried and species became extinct because of the extensive degradation.

Both peasant uprisings during the Qing took place as population growth peaked. Although chaos from wars and famine eased population pressures, the numbers were still astronomical. Population growth in the late Qing was uncompromising in its effects on the environment. Class oppression and feudal exploitation induced bankruptcy and emigration by many peasants.

In the nineteenth and mid-twentieth centuries, imperialist powers invaded China and plundered the country's resources. The savage invasion by Japanese imperialists badly damaged forests, mineral reserves, and agricultural resources. The Japanese destroyed approximately 64 million m^3 of timber in Northeast China's primary forests. Industrial capitalism was late to develop in China, and therefore did not play a major role in environmental degradation until later in the century.

A brief survey of population and environmental issues in China for 4,000 years from ancient to modern times suggests the following conclusions:

1. Environmental degradation in China is due less to geologic history and more to the expansion of human activities.
2. The interaction of population growth and environmental affairs has a long history. The relationship is increasingly evident over time as agricultural technologies advanced and populations grew. A more indirect relationship exists between time and the environment's influence on population.
3. Socioeconomic and political causes for an increasingly degraded environment include the system of feudalism and government-sponsored population policies. Feudal production suffocated productive forces, and population policies generally encouraged increased birthrates.
4. Environmental degradation has coincided with land reclamation and imperialist attacks in addition to population growth.
5. Limited knowledge of environmental sciences and natural resource management has further aggravated environmental degradation.

Notes

1. Ehrlich, Paul.
2 Jian Bozan. "An Outline of Chinese History."
3. Tong Lizhong and Zhang Weiguo, "Artificial Desert West to the Bend of the Yellow River." *Encyclopedia.*
4 Shi Nianhai. "The Loess Plateau and Changes in Distribution of Agriculture, Forestry, and Animal Husbandry." *Histotry and Geography.*
5. "The Era of Emperors."
6. Wu Shengyuan. *History of the Chinese Population.*

4

Population and Environmental Issues in Contemporary China

Understanding the dynamic relationship between population levels and environmental affairs in China necessitates a comprehensive overview of contemporary population growth patterns and environmental problems since 1949. An examination of the internal mechanisms of this relationship may help to resolve the controversy on how best to regulate domestic population and environmental affairs.

Population Trends

Large-scale, high-speed industrial construction followed the founding of the People's Republic of China in 1949. Agricultural production also grew rapidly. As urban and rural construction, health care, and science progressed, people's living conditions improved and material accumulation grew. Two changes in population trends ensued. During the 1950s and 1960s, death rates decreased and natural growth rates increased, but birthrates did not fall until after the 1970s. Present population growth, nevertheless, remains high. General neglect and poor population policies in the 1950s and 1960s have been the overriding causes for China's currently desperate situation.

Rapid Growth

In 1949, the population of China totaled 540 million people and accounted for one-fourth of the world's population. A stable social environment and economic incentives to propagate have caused even more rapid growth rates since that time. According to the fourth national census in China, the population as of July 1, 1990, reached 1.16 billion—2.1 times that in 1949—accounting for 21.9% of the world's population.

Aggregated birthrates, one of the major indices for controlling population growth, refer to the sum of the ratios of the number of children born to women in a certain age group. The aggregated birthrate determines population birthrates over long periods of time. Before birth-control policies were implemented in 1973, the aggregated birthrate of Chinese women exceeded 3.0. After strict birth-control policies, rates fell as follows: 2.61 (1981), 2.42 (1983), 2.59 (1987), and 2.50 (1992). Although China's birthrate is low relative to that of other developing countries, it far exceeds that of most developed countries. A historically high aggregated birthrate is the main cause for high population growth in China.

Fluctuations

Population growth trends since 1949 can be divided into five stages. Figure 4.1 illustrates the relationships among birth, death, and natural growth rates.

The first major period of population growth occurred from 1949 to 1957. An end to warfare and war-induced poverty lowered death rates and contributed to this growth. In 1949, 20 people per 1,000 died and by 1957 the rate dropped to 10.8 per 1,000. The period also witnessed a major increase in live births relative to 1949. Birthrates from 1949 to 1954 averaged 37 per 1,000; from 1955 to 1957 the birthrate exceeded 30 per 1,000. The net population increase during this first stage totaled 105 million.

From 1958 to 1961, China witnessed a major slump in population growth. During this stage, the national economy suffered serious setbacks and living standards fell sharply. Natural disasters caused sharp decreases in birthrates and increases in death rates. As a result, natural population growth fell. In 1960, death rates reached as high as 25.4 per 1,000 and infant death rates totaled 330 per 1,000.

Figure 4.1 Birth (BR), Death (DR), and Natural Growth (NGR) Rates by Year, 1952–1984 (per thousand population)

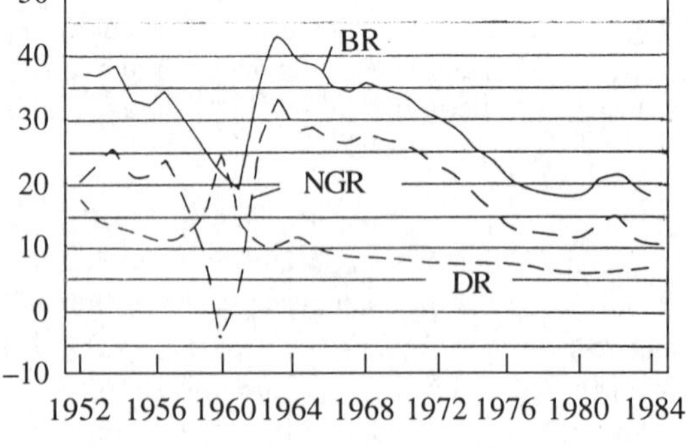

From 1962 to 1972, China experienced a second baby boom. Political movements such as the Great Leap Forward and "Communism Overheat" dissipated, and the economy began to recover. In 1962, the birthrate reached a high of 37.01 per 1,000. Birth-control policies during the Cultural Revolution (1966–1976) did not exist; birthrates fell, however, because of the general chaos to approximately 30 per 1,000. During the same period medical technology improved and death rates remained low. Natural population growth stayed fairly high at 26 per 1,000. Annual newborns totaled 2.5 million and net population growth amounted to 200 million.

From 1973 to 1984, population growth began to come under control. During this fourth stage, people became increasingly aware of the population problem. The Communist party and the government began to take measures that encouraged family planning. The population growth rate fell as follows: 30 per 1,000 (1973), 23.13 per 1,000 (1975), 20.91 per 1,000 (1981), and 10.81 per 1,000 (1984). This 12-year stage witnessed a net population increase of 140 million, the lowest of all of the stages since 1949. Family-planning initiatives have had many positive socioeconomic impacts, and have earned respect from other countries with similar problems.

In 1985, the natural population growth rate began to rise. Family-planning policies slackened; inappropriate loopholes for special interests such as farming naturally encouraged increasing birthrates. Beyond this, an expanding unemployed floating population also made the implementation of family-planning measures increasingly difficult. The nationwide responsibility system allocated a good part of implementation and enforcement responsibilities to decentralized neighborhood units. In 1984, natural population growth rates were 10.81 per 1,000. In 1987, these rates increased to 16.16 per 1,000 and have since steadied at 14 per 1,000. According to the fourth national census, the natural population growth rate from July 1, 1989, to June 30, 1990, totaled 14.7 per 1,000.

Population growth rates have fluctuated according to population policies. When population problems surfaced, policies were implemented and took immediate effect. China's frequent variations in population growth rates are rare in history. On the one hand, the fluctuation reflects the influence of birth-control policies. On the other hand, it indicates that certain policies on population control are commonly misunderstood by the public.

Young But Aging

Between 1949 and the early 1970s, China's population grew rapidly. High birthrates and low death rates have created an increasingly large and young population. An aging trend has begun since family planning took effect in the early 1970s.

In 1953, China's first population census indicated that proportions of the elderly and the young as a percentage of total population were 4.4% and 36.3%, respectively. After ten years of economic development, health-care improvements, and poor enforcement of family planning, birthrates increased rapidly. China's currently young population age structure is a consequence of this increase.

By the second census in 1964, the proportion of elderly people in the population fell even further to 3.5%, and the proportion of young people rose to 40.4%. The tremendous success of China's family-planning policies in the 1970s changed the population's age structure. According to a third census in 1982, the proportion of young people in the population fell to 33.6%, adults increased to 61.5%, and the elderly increased to 4.9%. At the time, the average age of the population was 22.9 years. In spite of China's young age structure, a more regular distribution is taking shape. After the 1980s, this tendency became more evident. A 1% sampling of the population in 1987 indicated that the proportion of young people fell to 28.7%; that of elderly people rose to 5.5%; and the average age increased to 24.2 years. According to the fourth national census on July 1, 1990, the proportion of young people in the population fell to 27.70%; that of elderly people increased to 8.59%, and the average age rose to 25.22 years.

Populations in rural areas are generally younger than urban populations. Birthrates in economically prosperous areas generally are lower and average ages are higher. In less-developed areas, birthrates are comparatively higher and age structures are younger. This reflects the phenomena of old age security and childbirth. If one is poor and lives in a rural setting, there is incentive to have a greater number of children to insure care and companionship.

As the national economy has improved, the average life expectancy in China has lengthened. China is soon to face a double-edged sword with regard to population problems. On the one hand, an age structure that moves steadily toward the elderly age classes leads to a top-heavy age pyramid of older, less productive citizens. On the other hand, the currently young population age structure presents problems of unemployment and social instability. These very different problems escape simple solutions.

Rural Versus Urban Trends

China's agricultural sector accounts for 80% of the whole population. This percentage has not changed much in spite of industrialization. Urbanization, however, has increased rapidly relative to China's socioeconomic status. Levels of urbanization often depend on the following:

1. Level of industrialization—how much of the labor force may be transferred from agriculture to other sectors.
2. Agricultural capacity—how efficient and productive the agricultural community is relative to urban demand.
3. National investment—how much money is spent on urban construction, infrastructure, and housing.

Present productivity does not match the fast rates of urbanization. Since the 1980s, China's urban population has increased sharply. Employment, urban infrastructure, and environmental protection have all suffered as a consequence. Efforts might focus on developing small- and medium-sized cities to alleviate pressures on large urban centers. Hong Kong's "satellite towns" in the New Territories exemplify how this might work. If rural inhabitants are at all influenced by mass campaigns that advocate "Leave farmland but not hometown," then a more diversified agricultural economy might take shape. Ideally, surplus labor from agriculture will find work in nearby towns.

Population Distribution: East Versus West

One of the most densely populated countries in the world, China has a population density of 118 people per km^2, as estimated by the 1990 national census. This is three times the world average. Greater densities exist in the Southeast and along the coast; more sparse areas are found in the northwestern regions of the country.

Six provinces or regions in the West—Inner Mongolia, Ningxia, Xinjiang, Tibet, Gansu, and Qinghai—contain 5.22 million km^2, accounting for 54% of China's total territory. The other 24 provinces, regions, and municipalities cover an area of 4.38 million km^2 and account for 46% of China's territory. The 1990 census reports that the population in the West, however, accounts for only 6.2% of the total. The region's density is a mere 14 people per km^2.

In comparison, the population of the 24 southeastern provinces accounts for 93.8% of the total population and has a density of 142 people per km^2. Densities of some well-known areas are as follows: Jiangsu Province and Shanghai, 738 people per km^2; Shandong and Henan provinces, 525 people per km^2; Zhejiang and Anhui provinces, 405 people per km^2; and Beijing, Tianjin, and Hebei provinces, 347 people per km^2. The more sparsely populated areas in the West might have population densities as follows: Heilongjiang Province, 78 people per km^2; Gansu Province, 49 people per km^2; Inner Mongolia, 16 people per km^2; and Qinghai, Tibet, and Xinjiang provinces, 6 people per km^2.

Regional differences in socioeconomic development and population distribution unquestionably lead to diverse and complex environmental dilemmas. Environmental policies made in Beijing need to conform to these location-specific differences.

Living Standards

At the founding of the People's Republic of China, living standards throughout China were low. In 1949, life expectancy averaged 35 years, one of the lowest in the world. Rapid socioeconomic development, however, greatly improved this and other health-care indicators. Infant death rates fell from 200 per 1,000 in 1949 to 80 per 1,000 in 1959 and to 35 per 1,000 in 1981. In 1985, infant death rates averaged 14 per 1,000 in urban areas and 35 per 1,000 in rural areas.

Higher levels of socioeconomic development in urban centers produce even more favorable results. By 1957, life expectancy increased to 57 years. From 1981 to 1985, life expectancies rose from 67.88 to 68.92 years. A 1% sampling of the nation in 1987 estimated life expectancy to be 69.05 years. Differences according to gender are worthy of note as well. Men generally lived an average 67.30 years, whereas women lived to an age of 70.66 years. The continual extension of life expectancy in China marks one of the country's many great achievements.

With regard to advances in culture, literacy rates since 1949 have considerably improved. People with above primary-level schooling have steadily increased in number as follows: 24.9% (1949), 34.7% (1960), and 60.4% (1982). According to the fourth census in 1990, university-educated students increased from 615 to 1,422 per 100,000 people. People who had reached their senior year in high school increased from 6,779 to 8,039 per 100,000; and, those at primary school increased from 35,237 to 37,057 per 100,000. Since 1982, the number of graduates at the university level have totaled 6.77 million. Of these, general university graduates number 3.45 million; and, adult graduates from television universities, training schools, night schools, correspondence schools, and employee training schools number 3.32 million. The number of graduates from middle school increased to 3.89 million people. A fourth census in 1990 indicates that the number of illiterates nationwide totaled 180 million, down by 50 million since 1982. The lower illiteracy rate, down 7.7% from 23.6% in 1982 to 15.9% in 1990 (Table 4.1), reflects extraordinary educational achievements.

Although living standards in China have greatly improved, more efforts need to be channeled to basic education. A large proportion of illiterates are youth; 2–3% of school-age children do not attend school. The population at or above a junior high school education accounts for 33% of the entire population; a large difference exists between this figure and that of developed countries. Because an educated public has a profound influence on socioeconomic development, improving school attendance is a primary objective in China.

Minorities

The number of minorities in China totals approximately 55. The predominant majority is the Han. Minorities with populations that exceed 1 million include the Zhuang, Hui, Uighur, Yi, Miao, Tibetan, Mongolian, Buyi, Korean, Dong, Hani, Bai, Tujia, Yao, Kazak, Dai, and Li. Minorities under 10,000 include Gaoshan, Tatar, Dulong, Elunchun, Hoche, Menba, and Luba. Different nationalities in China contribute to a wide range of diversity in culture, politics, and economics. In recent years, the population of minorities has grown considerably. From 1982 to 1990, Han people grew by 101.6 million, with an annual

Table 4.1 Illiteracy Rates in China, 1950–1990

Year	Illiteracy Rate (%)
1950	80
1964	38.1
1982	23.6
1987	20.6
1990	15.9

average growth rate of 1.29% per year. The population of minorities grew by 23.9 million during this same period, with an annual growth rate of 3.86%. As a percentage of the whole, minorities have risen from 6.67% in 1982 to 8.04% in 1990.

This relatively rapid growth may be attributed to many reasons. First, China has maintained a policy of equality, unity, and common prosperity for all minorities. Economic strengthening and social development have been encouraged in minority regions. Moreover, improvements in health and medical care have lowered death rates and lengthened life expectancies. Second, the government exempted minorities from many of the strict family-planning policies that impact the Han people. The government's family-planning policies have had a far greater impact on the Han people than on minority groups since the 1980s. Total birthrates for minorities have exceeded those of the Han. Third, many minorities who initially refrained from identifying their origin have decided to declare their true national identities. At the same time, children of intermarriages tend to choose their minority nationality rather than a Han identity.

Minority population growth should match economic development levels in their region. Every minority should make an effort to control its population for the sake of better living standards for all. This will further contribute to the diversity and prosperity of the Chinese nation as a whole.

Environmental Trends

With rapid economic development and sharp increases in population, environmental problems are increasingly evident. Pollution and environmental degradation have, in turn, affected daily productivity and national economic development. A brief survey of the historical context of current environmental affairs would best serve an analysis of current trends.

Historical context commonly explains the type, degree, range, and evolution of environmental affairs. Five unique conditions characterize China's past and present environmental trends: (1) stage of economic development, (2) agriculture-based economy, (3) resource availability, (4) technological development, and, especially, (5) vulnerable ecosystems. First, in spite of considerable progress, the national economy still lacks sufficient strength to solve many environmental problems. Second, a population predominantly supported by agricultural production has had a major impact on the environment. Third, China benefits from its vast territory and rich resources; it is plagued, however, by resource shortages, poor distribution, and pollution from coal burning. Fourth, advanced pollution control technologies exist, but are limited and useful only insofar as public concern for the environment exists. Lastly, China's environment is diverse. Current decisionmaking mechanisms and priorities weigh in favor of economic development, however, when the environment and the economy are at odds. The unique background of China's environmental affairs needs to be addressed before contemplating solutions.

A vast land, a large population, varied climatic conditions, and uneven economic development characterize China. Hence, the country's environmental problems are complex and varied: water pollution, atmospheric contamination, solid waste disposal problems, noise pollution, excessive deforestation, grasslands degradation, soil erosion, and habitat destruction. The problems are commonly interrelated and always complex.

Water Pollution

Water pollution in China is caused mainly by drainage from industrial wastewater. Pollutants affect both surface and groundwater. They commonly include oxygen-consuming substances and phenols. Exhausted industrial wastewater in 1990 amounted to 2.9 billion tons. The heavy metals—mercury, cadmium, lead, and chromium (VI)—account for 2,198 tons; arsenic accounts for 1,226 tons; cyanides account for 3,923 tons; phenols account for 9,325 tons; and petroleum residuals account for 66,560 tons. Since the 1980s, the quantity and quality of pollutants have changed.

Sections of rivers passing through cities are commonly very seriously polluted. National statistics indicate that 7.7% of rivers that tolerate waste exceed national standards. Large rivers are generally lower in quality than small rivers. Suspended sediment in the Yangtze River exceeds national standards, as do suspended-sediment levels in the Yellow River. Concentrations of mercury phenols and petroleum by-products all exceed set standards to varying degrees. Some sections of the Pearl River also have pollution levels of oxygen-consuming substances and nitrates that exceed national standards. The Huaihe and Songhuajiang rivers have been badly polluted by phenols, nitrates, lead, and cadmium. Annual average values of oxygen-consuming substances in the Liaohe River also exceed standards.

Soil nutrient loss into lakes and rivers is another common problem. Research on 34 major lakes across China indicates that 44.1% have a low to medium nutrient content; 29.4% have a medium to high nutrient content; and 26.3% have an excessive nutrient content. If no effective measures are taken, water contamination by soil erosion and nutrient runoff will worsen.

Coastal marine pollution occurs at varying degrees. Water quality near coasts, river mouths, and bays tends to be poor. Approximately 10% of Jiaozhou Bay is now affected by water pollutants. Bottom pollution is so serious that species diversity has fallen drastically. Nutrient salts and petroleum materials commonly exceed standard data. Nutrient salts at the Pearl River Delta now exceed water quality standards by a factor of 3. Algae growth increasingly afflicts coastal areas. In 1990, China's coast was plagued by 34 such excessively nutrient-rich areas.

Underground water quality near cities has begun to deteriorate. Main pollutants include nitrates, sulfates, phenols, arsenic, cyanide, and chromium. Overuse has lowered water tables as well as surface flow. In Hebei Province, for example, water levels have fallen at an annual average rate of 1–2 m. To date, 30 underground water funnels have formed.

Increasing amounts of water are polluted. Research on 99 water sources in 44 cities found that water quality at 51 sources failed to meet standards. Major pollutants include colibacillus bacteria and oxygen-consuming substances. Pollution of water sources not only has aggravated the already tight demand for a potable water supply, it has had a serious impact on people's health.

Air Pollution

Air pollution in China is mainly a result of coal consumption. Primary pollutants include dust and sulfur dioxide (SO_2). Since 1981, quantities of dust and gas emissions have risen more or less steadily. Winter and spring pollution levels generally exceed those in the summer and fall. Moreover, northern cities generally pollute more than southern cities. Pollutants vary, however, in that smoke and dust are more of a problem in the North than in the South, where sulfur dioxide is the primary pollutant. Daily and annual averages of micro-suspended SO_2 and N_2-O_2 molecules exceed set standards (Table 4.2).

According to government research, daily average values of micro-suspended grains for all cities nationwide in 1990 totaled 387 $\mu g/m^3$. Approximately 475 $\mu g/m^3$ occurred in northern cities and 268 $\mu g/m^3$ occurred in southern cities. The daily average of SO_2 is 94 $\mu g/m^3$ for all cities; the value accounts for 80 $\mu g/m^3$ in northern cities and 112 $\mu g/m^3$ in southern cities. Daily averages of the N_2-O_2 compound are 42 $\mu g/m^3$ for all cities; the values are 47 $\mu g/m^3$ for cities in the North and 38 $\mu g/m^3$ for cities in the South.

Atmospheric pollution in China currently mirrors levels in developed countries during the 1950s. A country's level of industrial development will largely determine the extent of atmospheric pollution. Concentrations of suspended particles and SO_2 in China have reached values that parallel those during the London smog accident. Unusually cold winters, which require additional heating for homes, add to the pollution in China's major industrial cities. Pollution levels in these cities are commonly much higher than current levels in London or Tokyo.

A by-product of the atmospheric pollution, acid rain has recently become increasingly severe. According to an analysis of rainfall in 23 provinces, regions, and municipalities, 44.5% of the samples detected acid rain. Cities where acid rain is detected in half of total rainfall samples include, among others, Qingdao, Suzhou, Chongqing, Guilin, Changzhou, Yibin, Fuzhou, Hangzhou, Guiyang, Guangzhou, Nanchang, Duyun, Changsha, and Nanjing. Average pH values of

Table 4.2 Content of Primary Pollutants in China's Cities, 1986–1990 (μg/m³)

	Total Suspended Grains		SO_2		N_2-O_2 Compounds	
Year	Northern Cities	Southern Cities	Northern Cities	Southern Cities	Northern Cities	Southern Cities
1986	715	391	105	108	55	41
1987	690	443	96	125	51	45
1988	674	434	89	101	48	42
1989	526	318	93	120	51	43
1990	475	268	80	112	47	38

rain are shown in Table 4.3. The analysis attests to the regional patterns of acid rain deposition in China. Special Economic Zones (SEZs) that foster industrial development unfortunately also generate high levels of acid rain. Acid rain is heaviest in the Southwest, where monthly averages fall below pH 5.

Solid Waste Disposal

In 1990, China produced 580 million tons of solid industrial waste, 10 million of which ended up in China's river systems. This waste would occupy a total ground area of 59.7 ha. In recent years, the stockpiling of industrial waste has rapidly increased. Nonindustrial waste has risen with population levels. If each person were to generate 1 kg/day of material waste and 1 kg/day of excrement, the annual quantity of these wastes in China would total 146 million tons. Most waste is neither treated nor recycled. The utilization rate of industrial wastes amounts to 26%; urban waste treatment is less than 6%; and the transportation capacity for solid wastes suffices for only 60% of the total amount. As a result, a large amount of untreated, recyclable waste forms stockpiles in city suburbs and pollutes rivers and streams.

In recent years, solid waste pollution has become increasingly serious. The waste is commonly flammable, toxic, and generally dangerous to any environment. More than 150 pollution accidents have been reported in Beijing, Shenyang, Kunming, Nanjing, Xi'an, Chengdu, Zhengzhou, and Hangzhou, among other cities. This has had an adverse effect on industrial and agricultural production, as well as on public health.

Noise Pollution

Noise pollution is commonly a function of economic construction; it can impact people's health and daily productivity. In 1990, noise within cities was attributed to the following: industry (26.7%), traffic (32.7%), and other routine sources (40.6%). Approximately 94% of major cities' traffic noises and 67% of medium-sized cities' noises exceeded set standards of 70 dB. Noise in city centers commonly exceeded the allowable 50 dB standard for cultural and residential areas.

Table 4.3 Average pH Value of Rain in China's Major Cities

City	pH value
Guiyang	4.07
Chongqing	4.14
Changsha	4.30
Nanjing	4.59
Hangzhou	4.72
Yibin	4.87
Shijiazhuang	5.36
Wuhan	5.47
Beijing	5.96
Tianjin	5.96
Jinan	6.10

Noise in China's cities is characteristically high. Approximately 66% of road sections have traffic noises that exceed the 70 dB standard. With respect to cities' overall noise, two-thirds of the sampled areas exceeded 55 dB. In recent years, the number of vehicles in cities has increased rapidly. City construction and traffic administration have not allowed for these increases. Traffic congestion and noise therefore has risen steadily. Due to increases in urban construction, specifically infrastructure, high noise areas are now extending into suburbs and peripheral towns.

Deforestation

According to the third national forest survey, China's forested areas cover 125 million ha. This amounts to a standing wood volume of 9.14 billion m^3 and a coverage rate of 12.98%. Relative to other countries, China is poor in forest resources, ranking 131st by percent of forest coverage. China's forested area per capita is 0.115 ha; standing volume per capita is 9.1 m^3. These figures are much less than the world averages of 0.65 ha for forest coverage per capita and 72 m^3 for standing volume per capita. Once rich in forest resources, poor management in the early part of the twentieth century set the country far behind. Short-term gains before 1949 have caused long-term costs.

Since 1949, China has launched large-scale tree planting campaigns. A low rate of preservation, poor forest management, and natural pathogens have reduced forest coverage and standing volume. Government statistics indicate that forest coverage from 1978 to 1991 was below 12%. In recent years, however, there has been an increase in forest coverage because of efforts at all levels. Age structures have changed, however, in that young forests have expanded in place of older forests.

Without urgent measures, harvesting operations will be severely limited within 5 to 6 years. Unsustainable forest management has disastrous consequences—wood shortages, altered ecosystems, species extinctions, soil erosion, riverbed deposits, water logging, and drought.

Grasslands Degradation

Moreover, China's grasslands are increasingly threatened by anthropogenic pollution and poor management. Grasslands in China cover 312 million ha, of which 10.53 million ha are artificial. Due to long-standing neglect and poor use, their niche in the environment is increasingly fragile. Cultivated grasslands amount to an estimated 7.2 million ha. Approximately one-third of this land has lost its reproductive capability and become desert. Due to excessive grazing, grasslands were destroyed at a rate of 15% per year in the 1970s. This number increased to 30% in the mid-1980s. Grassland destruction has had adverse effects on an industry that contributes to its demise, animal husbandry. Although professionals have taken some measures to protect and improve them, grasslands are degenerating and becoming less productive throughout China.

The total area of arable land in China has sharply decreased. At present arable land amounts to only 98 million ha, or less than 0.087 ha per capita. This is one-third the world average. In the early 1950s, farmers' initiatives and economic development placed agriculture as a priority. Hence, arable land masses increased. Since 1958, however, the rapid growth of industry, transportation, and urban construction has steadily eroded this land mass. Presuming that new arable land areas offset exhausted arable land areas, the annual net reduction of arable land is about 144,000 ha, bringing the total net reduction of arable land after 30 years to 4.32 million ha. This decrease in arable land has lowered the potential for agricultural productivity and put existing arable land areas under great pressure. Soil infertility and nutrient runoff from the misuse of fertilizers are only two of the disastrous symptoms of this pressure.

Soil Erosion

Soil erosion is an additional problem in need of attention. Incessant deforestation of sloping landscapes, excessive timber harvesting, and overgrazing have robbed some areas of their fertility. In the early 1950s, soil erosion affected 1.16 million km^2. At present, erosion has impacted 1.5 million km^2. Annual erosion amounts to approximately 5 billion tons, or 17% of the world's total soil erosion. The Yangtze River and Loess Plateau are two areas of severe soil erosion. The annual amount of mud and sand that flows into the sea from the Yangtze drainage area is 2.4 billion tons, nearly half of the total drainage in China. Soil erosion from the Loess Plateau affects 430,000 km^2, or 79% of the total plateau. Annual amounts of mud and sand drained into the sea from the Yellow River add up to 1.6 million tons. The Yellow River ranks highest in the world for the amount of suspended particles in the water. Once the cradle of Chinese civilization, the middle sections of the river are now barren hills and trenches.

Heavy soil erosion has caused the deposition of increasing amounts of sand and mud into the Yellow River channel. To prevent floods, dikes have to be raised. At present, the riverbed at some sections of the river is 10 m high. When floods arrive, dikes break and threaten surrounding areas. Annual soil loss is several million tons; in some regions, layers of soil 0.2 to 1 cm thick are washed away. Formation of such soils by natural processes may have taken from 100 to 400 years. The erosion not only damages soil fertility but also suffocates rivers and lakes. Approximately 22 of China's major reservoirs are heavily deposited with sand and mud. In 7 reservoirs built along the Yellow River, deposition has reached 40–75%. The economic losses from soil erosion add up to RMB 30 billion yuan per year.

As soil fertility erodes, sands predominate. In the North, 1.49 million km^2, or 15.5% of China's total land area, is desert. Deserted land amounts to 176,000 km^2 and potential desertification afflicts 158,000 km^2. Approximately 56,000 km^2 more of land has become desert since the beginning of the century. From the 1950s to the 1970s, desertification has increased at an annual average rate of 1,560 km^2. Since the beginning of the 1980s, however, desertification has slowed. Approximately 367,000 ha of land are still converted into desert every year, however. At present, desertification affects 3.9 million ha of farmland, 4.9 million ha of grazing lands, and 2,000 km of railway.

Reduction of soil fertility and soil salinization are common symptoms of desertification. Poor irrigation has aided secondary salinization of soils by 40%, from 4.67 to 7.8 million ha. Soil salinization has increased in the Huanghe-Huaihe-Haihe Plains by 120%, from 1.9 to 4.1 million ha, and it has increased in Inner Mongolia's Hetao Plain by 130%, from 44,000 to 105,000 ha. The second national general survey of soils surveyed 1,403 counties. The counties' soil conditions indicate that the average organic content of arable lands was only 1.5%. Approximately 59% of the arable land is poor in phosphorus; 23% is poor in potassium; 14% is poor in both phosphorus and potassium; and 86.7% has nutrient deficiencies of some sort.

Habitat Destruction

Deforestation and degradation of grasslands have threatened the habitats of many plants and animals. Habitat destruction is a leading cause of species extinction. Since the end of the nineteenth century, China's many ecosystems have been damaged more than most other nations' ecosystems. Approximately 15% to 20% of various vegetation types are threatened by extinction. Documents collected for 30 years indicate that wild animal habitat for the high-nosed antelope, wild camel, white-finned dolphin, wild elephant, hillside deer, golden monkey, giant panda, and Northeast tiger, among other species, is shrinking. According to the island biology theory of evolution, if 90% of the original inhabitants are destroyed, half of the vegetation will go extinct and the other half will continue to struggle for existence on small, crowded islands. These species may also face extinction because of natural calamity and specific requirements for pollination and dispersion. If no effective measures are taken, the range of animals and plants on the verge of extinction will widen.

Conclusion

In sum, 5 factors define China's current population and environmental problems: (1) a burgeoning population, (2) poor resource use by industry, (3) insufficient funding, (4) insufficient science and technology, and (5) dependency on coal as a primary energy source.

The Population Problem

Most environmental problems in China are related to recent sharp increases in population. This point will be addressed in both theoretical and practical terms in subsequent chapters.

Irresponsible Industry

China has long focused on increasing industrial output and has ignored the stewardship of its natural resources. High levels of consumption at poor efficiency have caused serious environmental degradation. To produce the same GNP, energy resources consumed by China have been 5.0, 4.7, and 3.1 times as high as amounts consumed in Japan, France, and the Federal Republic of Germany, respectively. Although a great deal of attention has focused on further improving China's economy, few efforts have focused specifically on efficiency.

Underestimating the value of natural resources in the national accounting system is a chief cause for waste. The status quo holds that, whereas final products are well priced and raw materials are moderately priced, natural resources have almost no price. Property rights issues are also commonly blurred. With little to no accounting for natural resources, national economic output values are grossly distorted and resources are continuously depleted for socioeconomic development.

Unsustainable use of natural resources is not only a technological problem, it is additionally a conceptual problem. Unless the current status quo fundamentally changes, it will be impossible to resolve many of China's current and future environmental problems.

The distribution of industry throughout the country has further contributed to environmental pollution and degradation. In the 1950s, China's economy was guided by mass campaigns and slogans such as "Go all out for steel and iron smelting." Small, localized iron and steel manufacturers sprang up everywhere from Beijing to rural farmers' backyard furnaces. Existing rules and regulations were broken, and locations for industry were chosen in wanton disregard. Another slogan, "Scatter about the mountains and enter caves," directed many factories to remote mountains and gorges where they subsequently discharged large amounts of pollution. If conditions for diffusion and dilution were poor, serious air and water pollution resulted. A third slogan promoted, "Change the consumption city into a production city." As a result, polluting industries were established in cultural city centers. With such scattered distribution, the construction of pollution prevention and control facilities is made difficult.

Strained Financial Resources

China's current environmental problems necessitate financing and long-term commitment. Economic development also mandates these two criteria. The ends are seemingly difficult to reconcile.

As an example of the enormous financial costs of pollution, the expense for secondary sewage treatment in cities all over the country would add up to a minimum RMB 30 billion yuan. Moreover, capital to control industrial and urban air pollution would amount to at least several hundred billion yuan. These figures do not include operational fees that almost double the amounts. Needless to say, such large investments exceed the country's current financial resources.

Although China spends roughly RMB 10 billion yuan on environmental protection annually, the amount and scale of this investment falls far short of real targets. Controlling environmental pollution and degradation within a country of 960 million km^2 of territory and a history of several thousand years mandates far greater financial attention.

Relative to other developed and developing regions of the world, China's investment as a percentage of GNP in environmental protection is at the low end of the scale (Table 4.4). Because of insufficient funding, legislation is difficult to implement and enforce.

Low Levels of Science and Technology

A principal factor of environmental degradation is the level of science and technology within a country. In conjunction with this is the human component—

Table 4.4 Percentage of GNP Allocated to Environmental Protection

Country	Period	Percentage of GNP
Average of the world	1970–1980	0.5–2
Developed countries	1970–1980	1–2
Developing countries	1970–1980	0.5–1
East Europe	1980	0.67
Africa	1980	0.2
Oceania	1980	0.7
United States	1970–1979	2
Great Britain	1970–1979	2.4
Austria, France	1970–1979	1.4
Japan	1972	1.8
Federal Germany	1975–1979	2.1
Belgium, Sweden	1974	1.0
China	1981–1988	0.65
	1990	0.7

the willingness to use available technologies and exploit resources less. Agricultural production in China stretches back several thousand years, but because of poor agricultural technology it is still very dependent on weather. In some remote areas, slash-and-burn cultivation still exists. Irrigation is still largely carried out by flooding, with utilization rates of only 20–30%. This not only wastes valuable water resources, it also causes soil salinization. Forest destruction by controllable pathogens is considerable because technologies used in forest pest management are not current.

Environmental protection industries and services have recently taken root. Bogus products, however, have unfortunately entered the market unregulated because of the healthy demand for them. This situation seriously affects the growth and development of environmental protection facilities.

Coal Dependency

Energy consumption in China demands 800 million tons of standard coal per year. China thus ranks first among coal-consuming nations (Table 4.5). Mineral energy resource use accounts for approximately 95% of all energy sources.

Among the various energy resources, coal affects the environment most seriously. Coal dependency is unfavorable. Compared with petroleum, natural gas, hydropower, and nuclear power, coal has devastating impacts on the environment. Smoke, dust, and other harmful discharges are the main culprits of air pollution. Moreover, residual wastes from coal use contribute to China's solid industrial waste problems. Every winter, cities in North China are draped in a veil of smoke and dust, harming daily productivity and people's livelihood.

Rural areas are commonly short of energy resources. Every year, peasants burn 400 million tons of plant material for their energy needs. The practice has a low heat efficiency, damages vegetation, and can harm the soil's organic layer—and these are only some of the side effects of energy shortages and overconsumption.

Table 4.5 Primary Energy Consumption as a Percentage of Total Energy Consumption

	World	China	United States	Japan	France
Petroleum	38.0	17.1	43.2	55.6	43.0
Natural gas	20.0	2.3	22.4	10.2	12.5
Coal	30.0	76.0	23.3	18.8	10.0
Nuclear power	5.0	0	6.1	10.3	32.8
Hydropower and others	7.0	4.6	5.0	4.9	1.7

5

Population and Land Availability

Environmental degradation and dwindling amounts of arable land have become serious problems in many countries and regions of the world. As populations expand, governments have increasingly paid close attention to land-use policies. It is China's turn to do the same.

Land is an essential component of the human environment. The world's varied land resources sustain all sorts of human activities. Land resources often determine a country's population characteristics. Three primary food sources include farmlands, grasslands, and oceans. Farmlands and grasslands provide 5.56×10^{15} calories per year; oceans provide 3.40×10^{14} calories per year. Food supplied by land amounts to 99.4% of the total. Land resources additionally determine government investments and the degree of capital construction.

Population distribution and migrations depend heavily on the distribution of land resources. A country's natural resource and socioeconomic conditions are, in turn, two major factors that influence population distribution. Densely populated areas typically occur in temperate and rainy areas with fertile soils. Conversely, sparse populations exist where excessive temperatures cause poor resource availability. In areas with relatively advanced socioeconomic development, populations are generally centralized. Moreover, population migrations are often a symptom of excessive land exploitation. New forms of resource exploitation, including mineral and resource extraction, may further contribute to land degradation and China's floating, unemployed population.

Poor land use has historically caused crises. Fluctuations in population levels and development rates further contribute to the complexity of land-use problems in China. Continuous advances in technology and increasing amounts of land under cultivation have drastically changed humankind's relationship with the land. Population increases and increased productivity have exacerbated land-use problems.

From 1000 A.D. to the 1970s, the world's population increased by a factor of 12. This increase has undoubtedly reduced the amount of land resources per

Figure 5.1 Population and Land Use

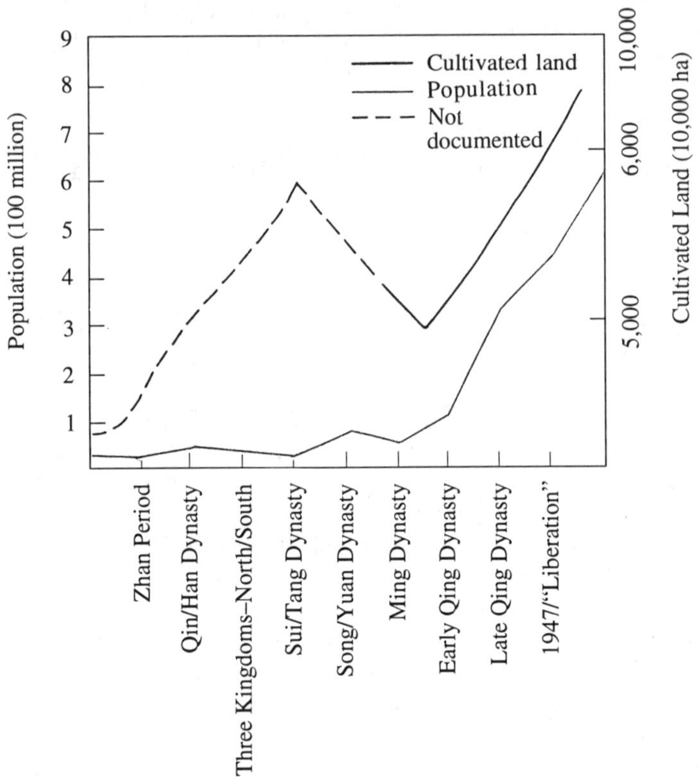

capita. As the human population threatens to exceed the land's carrying capacity, famines and other crises become more frequent. Sustainable socioeconomic development will occur only when population growth and land cultivation reach equilibrium. Such a harmonious, coordinated path of development depends on conscious government initiatives to control population growth and protect the environment. Scientific research should complement and enhance the efficiency of such land-use policies.

If population growth is neglected and efforts are not made to increase land utilization rates, conflicts between land-use alternatives will sharpen. Science and technology may begin to alleviate the conflicts. Sound population-control policies are a preventative, long-term approach to the problems.

In antiquity, humankind tapped the land's resources unmindful of future impacts on population policies or land-use principles. As populations increased, large tracts of unused land came under cultivation. Occasional drops in population alleviated pressure on the land so that areas that once were cultivated sat idle (Table 5.1).

Table 5.1 Population and Arable Land Growth Rates, 1949–1990

	1949–1952	1953–1957	1958–1965	1966–1978	1979–1984	1985–1990
Total annual population growth (%)	2.0	2.4	1.5	2.2	1.2	1.3
Annual increase of arable land (%)	+3.3	+0.7	+1.0	+0.9	+0.3	–0.5

Two large-scale migrations have occurred in China's history. From the East Han to the Sui dynasties (25 to 619 A.D.), cultivated land increased by nearly 40%. Cultivated land further increased during the South Song dynasty's reign (1127–1279 A.D.). Drops in population naturally increased the amount of per capita arable land. From the Ming dynasty to the Republic of China (1368–1949), the population grew steadily and arable lands concurrently expanded.

On the eve of the Communist victory in 1949, China's total arable land amounted to 97.3 million ha. Immediately thereafter, large-scale reclamation of unused lands infringed on land potentially used for capital construction. Cultivated land increased year after year. By 1957, cultivated areas expanded to 112 million ha. Expansion continued until the early 1980s, when rapid growth in other sectors of the economy caused the rate of agricultural expansion to recede (see Table 5.1).

Population growth has generally outstripped the expansion of cultivated areas. This tendency suggests that, in spite of China's vast territory, land availability has become an obstacle to economic growth. As both the economy and the population expand, land shortage problems and environmental degradation increase.

China's 9.6 million km^2 amount to one-fifteenth of the world's land mass and one-fourth of that of Asia. Grasslands account for the largest area. Forested areas and arable land follow second. Lands unsuited for agriculture include desert, mountains, and the tundra, which together occupy more than one-third of China's territory. Land used for farming, forestry, animal husbandry, fisheries, industry, and other sideline production amounts to 610 million ha, or 63.5% of the territory.

Annual biomass production in China amounts to 8.87×10^9 tons and accounts for 7.59% of the world's total output. This averages out to 861.45 tons per km^2. In arid and semiarid areas, biomass production is only 20.9% that of the country's total. These areas cover more than half of China's territory. The natural productive capacity of China's land resources per unit area is above the world average, but on a per capita basis, China's productive capacity is one-third that of the world average. China therefore needs to increase productivity on its available land resources while also implementing population control policies.

At present, China has a population of 1.13 billion people, or 22% of the world's total, but arable land amounts to less than 100 million ha, or 7% of the world's total. Per capita arable land, forest land, and grassland in China amount to 33%, 20%, and 25% of world per capita totals, respectively. Per capita cultivated land is only 0.11 ha, or one-thirty-fourth of the level in Australia, one-twentieth of the level in Canada, and one-ninth of the level in the United States. Among 26 countries with populations greater than 50 million, per capita arable land in China ranks third to last. Moreover, arable land is unevenly distributed. For example, per capita arable land is 0.05 ha in Zhejiang Province, 0.06 ha in Sichuan Province, 0.04 ha in Fujian Province, and 0.30 ha in Heilongjiang Province. As the population continues to grow, these figures are bound to fall. As pressures become more acute, the equilibria of countless ecosystems are jeopardized. Even with advanced science and technology, the land's productivity is limited.

More than 15 million babies are born annually in China. At the same time, cultivated areas fall by 162,000–202,500 ha per year. Per capita cultivated areas therefore fall by one-tenth every 3 to 5 years in most provinces. In Ningxia Hui Autonomous Region, for example, cultivated areas have fallen at an annual average rate of 13,000 ha since 1979. By 1990, 147,000 ha had been taken out of cultivation. In the meantime, population steadily increased at 3.4% per year. Relative to levels in the early 1950s, the population has increased by a factor of 2.5 and per capita arable land has decreased to one-third of previous levels.

Regions with the fastest rates of population growth are, unexpectedly, the arid and semiarid regions. They are typically backward, inaccessible, and plagued by natural calamity. Shanxi, Shaanxi, Gansu, Ningxia, Inner Mongolia, Qinghai, and Xinjiang provinces are several such examples, with population growth rates of 1.391% in Shanxi, 1.581% in Qinghai, and 1.828% in Xinjiang. Food availability is a primary concern, especially in circumstances with such high population growth rates. Since traditional farming techniques do not adequately increase productivity, peasants are compelled to cultivate undisturbed land. They often target marginal areas with poor site quality, resulting in soil erosion, desertification, and soil infertility. Natural calamities exacerbate the poor conditions, causing mud and rock flows, mountain torrents, and sandstorms.

Exploiting the world's resources in current fashion is bound to backfire. Engles has pointed out, "We should not be intoxicated with our success over nature. For each success, nature retaliates. Every success brings an initial expected result. Later, the same success may carry an unforeseen result that negates the first." Thus the following cycle is established: reclaim undisturbed land → destroy natural vegetation → degrade site quality and ecosystem equilibria → experience crop shortage → reclaim additional virgin land. The cycle has two disastrous consequences. First, it destroys forests, grasslands, and other unique ecosystems. Second, the increased demand for labor to clear fields and tend crops spurs additional population growth.

The vicious cycle influences not only local areas but also surrounding areas and lower reaches of river systems. The Loess Plateau is an area that has recently suffered from such a cycle. Once free of gullies and lush with vegetation, the plateau has felt the impact of population growth. Humans began to cultivate sloped areas so that 50% of arable areas currently occupy sloped land. Sloping fields aggravate soil erosion. The Yellow River contains increasing amounts of silt in its effluent. Below the Three Gorges Reservoir, the river discharges 1.6 billion tons of silt annually, raising the river bed and hence the potential for mass flooding. Floods are a detriment to people's livelihood and survival. Continuous flooding has contributed to poverty in certain areas.

With its history grounded in agriculture, China currently has limited land resources in preserves. Land preservation efforts in arable areas are even less common. According to research conducted by the State Planning Council's Comprehensive Investigation Committee of Natural Resources under the China Sciences Institute, there are 35 million ha of land in preserves, distributed in areas north of 35° North latitude. Less than half, 11.87 million ha, lends itself to cultivation. Approximately 10.67 million ha are in dry areas of the Northwest. The Northeast, as characterized by grassland areas of Inner Mongolia, is another major source of wasteland, occupying 16.5 million ha. Few large contiguous areas exist in the South. Approximately 3.93 million ha of wasteland is widely distributed.

Arable wasteland is characteristically far from urban areas, difficult to access, inhabited by minorities, and sparsely populated. In Xinjiang, there are 8.2 persons per km^2, and wasteland totals 10 million ha. In Inner Mongolia, the population density is 17 per km^2 and wasteland also totals 10 million ha. Population densities in Heilongjiang, Ningxia, and Yunnan are all under 80 per km^2, and wasteland exceeds 600,000 ha. Coastal areas in the Southeast and North have far higher population densities of 280–600 persons per km^2, excluding Beijing, Tianjin, and Shanghai. Inverse relationships between population density and wasteland is a foreseeable long-term reality.

Among land resources suitable for agriculture, 70% may be considered grasslands. Excessive logging, the construction of dikes, and other anthropogenic disturbances have aggravated soil erosion, intensified desertification, and destroyed unique ecosystems including grasslands. Agriculture, fisheries, and animal husbandry are only three of many competitors for land. Inhabitants of Xinjiang Province established farms in the 1950s and 1960s, opening up 1.02 million ha of previously undisturbed area. Natural vegetation was destroyed and 0.480 million ha of land has since been converted into desert. Wind and sandstorms have caused the fields to be covered and forced inhabitants off their property. Moreover, inadequate irrigation and drainage systems and the area's geochemistry have caused massive secondary salinization. Salinization has threatened approximately 440,000 ha, or 43% of the army's reclamation efforts.

In the Bijie area of Guizhou Province, the percentage of forest cover fell from 15% in 1957 to 5% in 1990. To establish commodity grain bases in the early 1960s, reclamation farms cultivated 200,000 ha of grasslands in the Hulunbeier Prefecture of Inner Mongolia. Regional patriotism fueled conflicts between the state-owned farms and the collectives concerning animal husbandry operations. Moreover, natural conditions in some areas were so poor that in less than three years, cultivation ceased on 177,000 ha, or 75%, of the reclaimed land.

In the Three-River Plain area, 2.7 million ha of undisturbed land was reclaimed from 1949 to 1990; average annual reclamation amounts to nearly 70,000 ha. Poor drainage on much of this area has induced secondary salinization. Bouts of drought and excessive rain in addition to the condition of the saline and alkaline soils have forced state-owned farms to cease cultivation on about 133,000 ha. A primary reason for such poor land-use policies has been overzealous goals to increase national grain output. Peasants have further cultivated undisturbed lands at random because of poor planning and ineffective regulations.

Demands from a large population for farm and agriculture products places undue pressures on arable land. Soil erosion, desertification, salinization, and pollution are four common problems. Natural calamities such as wind and sandstorms compound these existing problems.

Soil Erosion

Soil erosion in China has affected 1.5 million km^2. The amount of nitrogen, phosphate, and potassium washed away each year equals the amount contained in 40 million tons of chemical fertilizer, causing an annual loss of RMB 7.14 billion yuan. Arable land that suffers from soil erosion accounts for one-third of all arable land. According to computer simulations, existing arable land will be exhausted in 10–30 years on the Loess Plateau, 5–10 years in the southern hilly areas, and 2 years elsewhere. Because of soil erosion on the Loess Plateau, the Yellow River had 1,590 breaks in its banks in more than 2,500 years and approximately 50 noteworthy shifts in its course. Mountains with sparse grass, thin soils, and gullies attest to the soil erosion. In Shaanxi Province, a small-scale peasant economy is dominant. Extensive cultivation and backward production exemplify the slogan, "Reclaim one more sloping field and receive one more steamed bun." An increasing amount of sloping fields are characterized by thin soil and low crop yields. They commonly output a mere 0.75 tons per ha. In

Table 5.2 Leading Causes of Desertification

Anthropogenic Desertification Factor	Percentage of Desertified Land
Irrational reclamation	25.4
Excessive grazing	28.3
Lumbering of woods	31.8
Industrial, mine, and communication construction	0.7
Improper use of water resources	8.3
Total proportion	94.5

Sichuan Province, soil erosion has impacted 380,000 km^2 with increases in droughts as well as floods.

Desertification

The sharp rise in population has induced farmers to wantonly reclaim undisturbed land to graze domesticated livestock. The resulting destruction of vegetation leads to desertification. Wind facilitates the process, combining dispersed barren areas into large contiguous deserts. Desertified land has grown to 333,000 km^2 in the North and 18,000 km^2 in the South. About 1,560 km^2 of land become desertified annually. Anthropogenic disturbance is the leading cause for desertification (Table 5.2).

Desertification is exacerbated by wind and sandstorms. In Xinjiang Province, for example, unchecked reclamation transformed 1,300 ha of farmland into desert. More than sixty households were forced to relocate in the process. In May 1986, a duststorm destroyed 90% of the cotton, wheat, and maize fields in the area.

Salinization

Approximately 10 million ha of once-arable, salinized land exist in China. Arable land threatened by salinization in four provinces—Xinjiang, Gansu, Ningxia, and Inner Mongolia—amounts to 30–40% of the total cultivated area. Major reasons for secondary salinization are as follows: irrigation raises groundwater and salt through capillary pores to the surface, flood irrigation rich in minerals brings salt to the surface, and salt layers are mixed in with the subsoil when land is reclaimed. Salinized land is more liable to deteriorate from drought and excessive precipitation. Gley horizons exist on 20% of all paddy fields and reduce rice yields considerably. Poor drainage and irrigation cause water to collect for long periods of time. Salinization and gley horizons then contribute to degradation. More rational, science-based land use will help to overcome many of these problems.

Urban Development

Population growth contributes to increases in housing, traffic, and other construction that necessitates land. Nearly 500,000 ha of cultivated land is impacted annually by one of three construction activities—state-sponsored, township-run, and private. The total reduction in arable land equals Fujian Province's cultivated area. Arable land taken out of production for urban construction is typically productive (grain yields of 1 ton per mu). This adds to a grand total deficit of 3.4 million tons of grain per year, enough to feed a population of 250,000.

Beijing is one of the more blatant urban areas that occupy potentially productive arable land. Cultivated areas within the city limits have dropped from 531,000 ha in 1949 to 419,000 ha in 1987. Outside of Beijing, the development of township enterprises has impacted land use in rural towns. In Shuozhou, Pinglou District, land used for township enterprises amounted to 500 ha by 1988. Peasant housing is another major factor in decreasing the land base for agriculture. In one county of Zhejiang Province, land used for building homes amounted to 530 ha. Land used nationwide for housing amounted to 100,000 ha in 1985 and 85,000 ha in 1986. In recent years, the tendency to use land for building houses has not changed. Urban growth, then, pays a heavy price in terms of reducing the nation's productive farmland.

Poor Land Management

Population growth and increasing demand for agricultural products have led to "predatory land management" in many areas. Land is used and abused in the short term without regard for future consequences. Currently, land with high soil fertility amounts to 27% of the total, land with average soil fertility amounts to 38%, and land with poor soil fertility amounts to 35%. Because incentives to invest in the land's fertility are not immediately apparent, site quality continues to deteriorate nationwide. At present, China's investment in soil fertility and productivity is one-fifth to one-tenth that of foreign countries. In the Siuhua areas of Heilonjiang Province, soil fertility was once excellent. Over several decades, however, soil fertility has plummeted. Organic content has fallen from 6% in the 1950s to 3% in 1990. Continuous plowing has made the soil hard and impervious. Although use of chemical fertilizers has risen from 60 kg per ha in 1960 to 318 kg per ha in 1980, the organic content of soils has consistently fallen. Investment in chemical fertilizer is an incomplete solution. Innovative land-use policies and use of organic fertilizers are additional approaches that might be pursued to reconstruct soil fertility and productivity.

Current Affairs

In spite of a vast resource base, China's per capita resource figures pale relative to those of other countries. The population continuously increases while the arable land base degrades. Land-use issues will determine the existence and development of the country. A comparison of trends in population control and land use might shed light on innovative policies that incorporate both concerns.

In 1990, China's population was 1.13 billion; if trends remain the same, this figure will grow to 1.28 billion by the year 2000. Government estimates forecast that the population will reach 1.64 billion by 2050. China's arable land resources are indirectly related to the growth in population. The 1990 figure of 99.33 million ha will fall to 96 million ha by the year 2000 and to 73 million ha by the year 2050. Per capita grain production in developed countries averages 600–700 kg per year. At a conservative per capita consumption of 500 kg of grain per year, China's population of 1.28 billion people will require 640 million tons of grain requiring an average per ha output of 6.7 tons. The unit output of grain, however, even in a highly productive year, is only 3.6 tons per ha. Self-sufficiency, then, will be met only if per capita grain output figures double. This is a daunting task. The current unit output of grain is at the world average, in spite of being half that of advanced, high-yielding countries. The potential productive capacity of China's arable land is 940 million tons of grain. The trends suggest that even if grain production were to be a primary focus, China will be compelled to import grain by the year 2000. To control the population and protect arable land is China's most pressing concern.

China's socialism mandates the development of industry and urban areas via national, collective, and peasants' residential construction. Meeting this objective will require development of 6 million ha of potentially arable land by the year 2000. At the same time, only 2.7 million ha of land may be opened to cultivation. Per capita arable land will therefore fall from the current 0.11 ha to 0.075 ha in 2000 and 0.02 ha in 2050. The conflicts that stem from an increasing population and a shrinking arable land base will only intensify.

Soil erosion rates manifest two different trends. If not controlled, soil erosion may impact 1.7 to 1.8 million km^2 by the year 2000, an increase of 13–20% from the present. The quantity of soil lost will increase by 20–25% and the land's productivity will fall. If effectively controlled, soil erosion may be limited to 1.3 million km^2, a reduction of 13% from current levels. Both environmental affairs and productivity would be spared from losses.

A similar situation exists with regard to land desertification. If current land use persists, human activities will intensify desertification. An estimated 75,800 km^2 may turn into sand dunes by the year 2000. At present, 10% of desertified land is under control and 12% is fairly improved. With effective measures, therefore, desertification may be alleviated. By the year 2000, reclaimed desert will amount to 26,000 km^2.

If land pollution is not effectively controlled, agriculture and animal husbandry will suffer. Food hygiene indices may fall to "medium-polluted" and vegetables may cause more harm than good to people's health. Death rates by pollution-induced diseases will rise. With effective controls, polluted land may decrease from the current 15% to about 7%; polluted grain may also fall from the current 80 billion kg to approximately 30 billion kg.

In sum, prospects for population control and land-use crises look dim unless effective measures are taken soon. Effective and comprehensive measures demand immediate attention and high priority.

Countermeasures

China's modernization drive has recklessly approached and often exceeded the carrying capacity of China's environment. Environmental impacts, namely pollution, attest to this fact. "Cherish, use, and protect every inch of cultivated land"—if China's peasants adhere to this guiding principle in addition to state family-planning and environmental-protection policies, China may steer clear of its present path toward resource depletion.

According to surveys by China's Academy of Science, the current carrying capacity of the land is 950 million people. As of 1990, this figure was exceeded by approximately 180 million people. The success of future policies will determine China's well-being as it enters the twenty-first century. Active policies to control population growth and protect arable land must be pursued.

Per Capita Land Base

A sustainable per capita land base needs to be more clearly defined. Then, solutions might include delimiting protected areas for farmland and prohibiting nonagricultural activity in these areas. An overall aim of land-use policies is to allow adequate land resources to all trades. Governments at all levels should orchestrate these policies. All individuals and collected interests need to cooperate. Those who defy cooperation should suffer punitive damages. National land administration departments need to implement a general program in light of current laws and regulations. To effectively protect arable land, these departments will have to research local conditions and conform measures to match each area's peculiarities. Lastly, duties as well as rights should be clearly defined and enforced in townships and villages throughout the country.

Marginal Farmland Areas

China's plight might be paraphrased as follows: "Poverty induces cultivation and cultivation induces poverty." Poor science and limited awareness of general ecological principles constitute a root cause for the present circumstance. Humans can either degrade or promote land productivity. To prevent degradation, public awareness needs to be strengthened through education and propaganda. People need to learn of the devastating effects of indiscriminate logging, slash-and-burn agriculture, and cultivation on steep slopes. Where soil imbalances currently exist, efforts should be made to leave the land fallow, prohibit cultivation, and grow grass and trees. On the contrary, where conditions call for cultivation, intensive use should be allowed.

Jilin Province has conducted a tremendously successful experiment in which agriculture and animal husbandry were delineated according to ecological models of the landscape. Socioeconomic and environmental benefits of the planning have been enormous. Per-hectare productivity from dry farmland has increased by 750 kg and income from timber and nontimber forest products has reached RMB 160 million yuan. Soil erosion has been brought under control and sandy areas stabilized. Unemployment is low and science and technology prosper.

Undisturbed Areas

China's stability rests in its ability to exploit undisturbed land resources. Of equal importance, however, is the country's ability to balance exploitation with protection and preservation. Rare plants and animals need to be protected in their unique natural environments. Ecosystems of all sorts serve more purposes than that of human subsistence. They act as a classroom, a laboratory, a pharmacy, a gymnasium, and a house of worship, among other things. Experts in resource ecology and technology should cooperate to comprehensively research the limits of exploitation. Preservation initiatives that balance amounts of indiscriminate exploitation and idle land need to be enacted.

Comprehensive development of agriculture, forestry, animal husbandry, and fisheries needs to replace a preoccupation with grain production. Socioeconomic and environmental factors are primary determinants of any such development plan.

Natural and Anthropogenic Disaster

In the past 40 years, China has made great strides in water and soil conservation. The progress, however, falls short of present rates of soil erosion. Resolute measures that integrate biology and engineering must be implemented to prevent and control pollution and erosion. Indiscriminate timber harvests and land conversion should be prohibited. In regions with relatively light degradation, water and soil conservation efforts should aim to retain runoff and increase water permeability. In regions with more serious erosion, hilly areas should be set aside as grasslands or forested areas. Field engineering should take steps to check encroaching desertification, thicken soil layers, retain soil moisture, and restore natural vegetation.

If prevention and curative measures are effectively implemented, land desertification in China might be controlled by the beginning of the next century. Historically poor land-use policies might benefit by focusing on where forestry and animal husbandry best fit local conditions. Interspersing zones that conform to the natural landscape appears to have worked for Jilin Province. Other aims might include restoring forests and grasslands on desertified farmland. Increasing forest coverage rates and protecting natural vegetation are additional priorities that may help to improve environmental affairs. Water conservation and agricultural measures might ameliorate secondary salinization, the buildup of gley horizons, and flooding. A comprehensive plan that controls river networks and drainage areas needs to be thought out and implemented. Water distribution might be allocated to farms in accordance with their salt-enduring characteristics.

Comprehensive measures to prevent land pollution include establishment of standards, controls on use of pesticides and fertilizers, suggested plowing procedures, and transplanting of soil. First, environmental policies, standards, and regulations need to be enforced at all levels of government. Second, research efforts should focus on developing effective pesticides with low toxicity and low residual effects. Use of chemical fertilizers should also be more tightly regulated and substituted whenever possible with organic fertilizers. Third, crop rotation and plowing should be improved to speed the decomposition of polluted substances. Lastly, soils should be transported to polluted areas to increase land productivity and dissipate pollutants.

Reclamation of Polluted Areas

Mining and extracting industries in general can seriously degrade land resources. If landscapes can be recultivated and improved, however, such industrial activity might be more tolerable. In the Tianchi Lake District of Yunnan Province, piles of residual from phosphorus mines have destroyed 4,700 ha of land. Natural vegetation in the area has been badly damaged and soil erosion has worsened. Precipitation in the area has caused a great quantity of the soil, fluorine, and phosphorus to empty into the lake. Since 1987, a comprehensive plan to control the degradation has had a tremendous effect. Land is recultivated nationwide at a rate of 2%. This lags far behind rates of 70–90% in other developed countries, however.

Controlling Urban Growth

Large urban areas occupy an increasing amount of China's naturally fertile land. If the expansion of urban areas is not controlled, surrounding arable land will dwindle. The establishment of small cities and townships should be encouraged in place of larger spreading urban areas. Moreover, urban construction should encourage tall buildings to increase population densities. Single-family homes that encircle yards consume more land per individual than do more centralized skyscrapers. This sort of urban life-style change is difficult to regulate, but would have a pronounced effect on the effective use of land resources.

6

Population and Forest Resources

Forests are a crucial component of many of the world's ecosystems. In addition to supplying timber and other forest products, forests have a myriad of functions, including carbon dioxide absorption, oxygen production, water conservation, barriers to wind and desertification, climate regulation, prevention of soil erosion, and preservation of wildlife habitat. Forested ecosystems are complex and active. Economics and monetary benefits often overshadow benefits that are not easily quantified. Consequently, indiscriminate logging has eaten away at China's standing forest reserves and the many functions they serve. A series of serious socioeconomic and environmental consequences are just now being felt. Forest degradation is one of the more prominent problems that China faces as the population and the economy expand.

Forest reserves provide the following four benefits: (1) fiber; (2) land; (3) wildlife, plants, and microorganisms; and (4) aesthetics and recreation. A forest will provide these benefits in one of two manners. It may be visible and direct, as in providing timber and nontimber forest by-products, or it may be invisible and indirect, as in the external benefits of oxygen production and water filtration.

As a source of timber, forests meet a worldwide timber demand of 3 billion m^3. This is mainly used for construction materials, papermaking, and other industrial uses. Nontimber forest products include oil, perfume, drugs, and rosin. In recent years, energy experts have begun to consider forests as a replenishable resource to exploit. The total value of forest products worldwide amounts to US $115.5 billion per year.

Forests, moreover, help shape regional climates, hydrologic cycles, and landscape aesthetics. Some defining features include wide geographic distribution and long rotation cycles. The ecological benefits of forests may, in fact, be 10 times greater than the more easily quantified monetary benefits. For example, a single hectare of forested land conserves 3,000 m^3 more water than does a non-forested area. Forest ecosystems are additionally home to a wide array of organisms and account for 90% of the total world output of plants. Trees are an important contributor of organic substances and an integral part of many organisms' food webs. Trees enrich plant and animal diversity. They might be thought of as a "gene bank" contributing to the variability of species. This characteristic alone contributes billions of U.S. dollars to agriculture, medicine, and industry every year.

Forests play an important role in conserving water and soil quality. If forest coverage, or basal areas, in mountainous areas exceeds 60%, a myriad of forest functions are brought into play. Fallen leaves and dead branches increase the organic content of soils. The various soil and rock substrates, in turn, help filter water as it makes its way to underground flows.

Forests additionally regulate microclimates. Foreign and domestic research has shown that microclimates may be regulated and disastrous weather systems prevented if the percentage of forest cover exceeds 30%. Forests also purify the atmosphere and help control air pollution. The content of carbon dioxide in the atmosphere increases sharply with industrialization and fossil fuel use. Each hectare of a broad-leafed forest consumes 1,000 kg of carbon dioxide and produces 730 kg of oxygen per day. Many industrialized countries face problems associated with dust in urban areas. Each hectare of forest can absorb 32 to 64 tons of dust. Moreover, forests purify air by absorbing harmful gases. One hectare of forest, for example, consumes 720 kg of sulfur dioxide annually. Forests additionally secrete agents that sterilize bacteria, fungi, and protozoa. This function makes forests an ideal place for plant regeneration. Forests may also insulate an area from noise pollution. Trees may reduce sound by 26 to 43 dB; trees along streets can reduce traffic noise by 8 to 10 dB.

The list of benefits continues, suggesting that trees are far more valuable than merely for the timber they supply. For instance, the Environmental Protection Research Institute of Jilin Province conducted a cost-benefit analysis of forest resource management options in the Changbaishan Nature Reserve that incorporated nonfiber forest benefits. They discovered that if they were to construct a reservoir large enough to conserve the same amount of water that healthy forest soils store, it would cost RMB 2.81 billion yuan. Preventing water loss and soil erosion by means of a terraced landscape would cost an additional RMB 2.97 billion yuan. Forests further prevent pest management costs by protecting wild birds that eat insects, common pathogen vectors. Along these lines, forests prevent damage in the Changbaishan area of approximately RMB 2.99 billion yuan. If one were to calculate the carbon dioxide absorbed and the oxygen supplied and further factor in the benefits of forest organisms and air purification, forests would increase in value by another RMB 480 million yuan. Not even accounting for the benefits of wind and sand control and climate regulation, the ecological benefits of the Changbaishan Nature Reserve alone amount to RMB 9.27 billion yuan.

Because forests have such important monetary and environmental benefits, careful forest management should receive high priority. Forested areas continuously decrease, however, with China's sharp increases in population. Forested areas once covered 7.6 billion ha worldwide. They now cover only 2.8 billion ha. Deforestation continues to increase at 20 million ha per year. If this trend continues, future generations might not know this world as we do today. Ecosystems of all kinds will be thrown out of equilibria and humans will face even more disastrous circumstances than we do at present.

Historically rich in forest resources, China now faces severe limitations. Population growth and an increased demand for arable land have rapidly depleted China's forests. The present percentage of forest cover is only 12.98%, lower than the world average and 120th among 160 countries and regions in the world. The low percentage of forest cover in China is caused mainly by a burgeoning population and a dwindling supply of arable land. Since the modes of agricultural production have not substantially changed, an increasing amount of land has been deforested by China's ever-expanding peasant-based economy.

A Brief History

During the Period of the Warring States, from the fifth to the third centuries B.C., forest coverage in China was 42.9%. Deforestation then increased at rapid rates for 5 principal reasons. First, population growth and increasing demand for arable land compelled peasants to carve out areas of the forest. Forests were one of the first areas targeted because soils were guaranteed to be fertile. Second, backward modes of agricultural production exhausted and destroyed large tracts of land. Third, feudal rulers engaged in large-scale construction, requiring vast tracts of forest for timber. The Great Wall and E'fang Palace are two such examples. Fourth, frequent warfare often entailed forest fires to resist the enemy. Lastly, feudal dynasties did not actively reforest areas.

After the Communists won in 1949, China began to deviate from this path of neglect. The government called people to arms with the slogan, "Plant trees to cover the country." This quickened steps to reforest barren mountainsides and riverbeds. The deficit in forested areas continued to rise, however, because of the demand for arable land and timber products.

Since 1949, there have been two large-scale incidents of forest destruction. The first occurred during the Great Leap Forward. Large tracts of forest were destroyed to make iron and steel. Areas that had no fossil fuel resorted to timber to heat furnaces for steel production. In addition, 60% of China's reservoirs were built during this period. The reservoirs flooded vast forested areas. In Chifeng, Inner Mongolia, the Hongshan Reservoir built in 1958 caused the loss of hundreds of thousands of pine trees.

The second large-scale destruction occurred during China's Cultural Revolution. University graduates were sent to remote areas in Inner Mongolia, Xinjiang, Heilongjiang, and Jilin to cultivate undisturbed areas and engage in other manual labor. In the 1970s, forest cover dropped to 12%. Since that time, coverage has rebounded to 12.5% in 1988 and 12.98% in 1990. Although this indicator has improved, other indicators of the health of the country's forests have not. Age structure and forest storage are two such indicators.

Current Affairs

Vast in territory and with a wide range of terrain and climate, China's forests include Northern Coniferous Forests, Mixed Coniferous and Deciduous Broad-Leaf Forests, Evergreen Broad-Leaf Forests, Tropical Monsoon Rain Forests, and Forest Steppe. China has a total of 8,000 varieties of plants, among which there are 2,800 tree species. Only 1,000 of these species constitute high-quality wood with any economic value. Diverse wildlife and plant species have helped to establish a unique gene bank.

Forest coverage is an important indicator of resource exploitation. Only a certain, well-distributed percentage of forest cover can help to improve agricultural production, add to the aesthetics of an area, purify the water and air, and prevent natural disaster. At present, forested areas in China amount to 125 million ha, or 12.98% of total land area, far below the world average of 30.67% (Table 6.1).

China's aggregate forest resources are among the world's largest, but the country lags in most per capita statistics. Per capita forest coverage is ranked 120th in the world at 0.11 ha, compared with the world average of 0.61 ha. Total reserves amount to 10.26 billion m^3, or 3% of the world average of 310 billion m^3. Per capita reserves amount to 9 m^3; the world average is 75.6 m^3, or 8.4 times China's figure.

Due to poor distribution of forests, a large population, and intensive logging, forested areas are now mainly distributed in areas that are difficult to access. The Northeast and Southeast make up one-third of the country's land mass, but

Table 6.1 Percentage Forest Cover and Per Capita Forest Area

Item	China	Finland	Korea	Japan	Sweden	United States	World Average
Percentage of forest cover	12.98	71	74	68	53	32	30.67
Per capita forest cover (ha)	0.11	4.98	0.91	0.22	2.87	1.34	0.93
Index of per capita forest area (China=1)	1	45.3	8.3	2.0	26.1	12.2	8.5

contain 43.0% and 65.9%, respectively, of the country's forests. The few forested areas in the Northwest and West are far from urban areas, and transportation costs are high. The average distance traveled to harvest timber is 1,200 km. Provinces vary tremendously in percentages of forest cover. Coverage exceeds 50% in some areas and drops below 1% in others. Forest cover in Fujian, Zhejiang, Heilongjiang, Jiangxi, Hunan, and Jilin is above 30%; coverage in Qinghai, Xinjiang, Shanghai, Ningxia, Tianjin, Jiangsu, and Gansu is below 4% (Table 6.2).

Forests denote nurseries, tree farms, natural woodlands, and shrublands. A common indicator that measures the degree of forest management within a country is the utilization ratio, which reveals the fraction of land with trees

Table 6.2 Population Density and Percentage Forest Cover in Provinces, Cities, and Regions of China

Province, City, or Region	Population Density (person/km^2)	Percentage of Forest Cover
Fujian	248	49.2
Zhejiang	407	39.6
Jiangxi	226	38.2
Heilongjiang	78	36.2
Jilin	132	36.1
Guangdong	353	34.0
Hunan	286	31.3
Yunnan	94	25.2
Hubei	290	24.2
Shaanxi	160	24.2
Guangxi	178	24.0
Liaoning	270	22.8
Guizhou	184	15.1
Anhui	404	13.5
Sichuan	188	13.3
Beijing	644	11.9
Henan	512	11.1
Hebei	325	10.6
Shandong	539	8.8
Inner Mongolia	18	8.7
Shanxi	184	7.3
Hainan	193	7.0
Tibet	1.8	5.3
Gansu	49	4.8
Jiangsu	654	3.4
Tianjin	777	2.7
Ahsnghai	2,118	1.7
Ningxia	90	0.9
Xinjiang	9	0.9
Qinghai	6	0.3
All of China	118	12.98

relative to the total amount of land set aside as "forests." In China, forests theoretically amount to 270 million ha, or 28% of the total territory. Only 125 million ha, however, constitutes land with trees. The utilization rate is therefore only 46.3% (125/270). Utilization rates in other countries are comparatively high: Germany (97%), United States (95%), and Japan (93%). China's low utilization rate suggests that its forests have been seriously depleted and poorly managed.

All too often regarded as stands of timber, forests in fact offer multiple other benefits. Like many other areas of the world, China is also guilty of managing its trees for timber alone. This is evidenced in the following figures regarding use: commercial forests (73.2%), economic forests (10.2%), shelter forests (9.1%), fuel forests (3.4%), bamboo forests (2.9%), and special-use forests (1.2%). Relative to commercial forests, other forest types are quite small and in need of protection.

Poor regeneration and excessive logging have caused poor growth rates. At present, average net growth nationwide is 2.66%. The annual production of wood is approximately 230 million m^3 and annual per capita production is 2.4 m^3 per ha.

Pressures on Forest Management

A low per capita wood reserve in China indicates that forest reserves are bearing too heavy a burden. After 1949, sharp increases in population increased the need for both timber and arable land. To meet the new needs, the state opened two large forest zones in the North and Southeast. These zones contain China's most valuable supply of timber. Timber output in the Northeast accounts for 33% of all forest reserves. A policy of harvesting remote areas and generally excessive logging caused forest resources in this area to dwindle. Among 140 forestry bureaus in the country, 61 harvested trees unsustainably and 26 depleted their reserves entirely. In Jilin and Heilongjiang provinces, forest reserves are nearly depleted. Only the Daxinganlin Mountain area retains part of its forest reserve. The forest fire of May 6, 1987, nevertheless, seriously damaged the stand. In the Southeast, forest reserves have already been depleted.

To meet local needs, actual cuts have far exceeded planned cuts. Planned and actual logging in fact differs by a factor of 2 to 3 in Fujian Province; by a factor of 3 in Heilongjiang, Jilin, and Guanxi provinces; by a factor of 5 in Guangdong Province; and by a factor of 11 in Yunnan Province. Consumption of wood has recently exceeded growth in volume by 25–50%. In Sichuan, Yunnan, Fujian, Guizhou, Jiangxi, Zhejiang, Hunan, Anhui, Guangdong, and Shanxi, the annual consumption of reserves has been 110.5 million m^3. In the early 1950s, forest cover was above 19% in Sichuan; it had fallen to 13.3% in 1990. In Yunnan, forest cover in 1990 was less than half of what it was in the late 1940s and early 1950s. In Xishuangbanna (Yunnan Province) from 1947 to

1980, forest cover fell 530,000 ha, from 57.8% to 30%. In Fujian from 1976 to 1980, forest coverage fell from 57.8% to 37.0%. In Heilongjiang, forested areas have dwindled in size by approximately 2.7 million ha since the early 1960s. In the 1950s, natural forests covered 8.6 million ha on Hainan Island. At present, the area is only 250,000 ha. Forest coverage fell from 25.7% to just over 7%.

Both the continual decline in forest reserves and the increases in demand for forest products have caused shortages in timber and other forest products. The annual consumption of wood adds to 200 million m^3. The average annual deficit in wood is 50–100 million m^3. In spite of increasing amounts of timber imports, the deficit still remains high. If the trend remains constant, forest reserves in China will be exhausted in 5 to 6 years.

One of the outstanding characteristics of China's population is that the majority live in the countryside. Rural populations have grown without interruption for quite some time. It follows that the demand for energy has also steadily increased. Still dependent on plant fuels such as firewood and crop straw, rural areas are increasingly short of energy. Without stable sources of energy, the population will find it increasingly difficult to carry on the daily activities of cooking and heating. The demand for large quantities of plant fuels is liable to induce widespread logging and general destruction of China's forest resources. In Jiangxi Province, more than 11 million tons of fuel wood is consumed annually. In Yichun, Heilongjiang Province, annual timber production is 5 million m^3. Approximately 60% of this amount is consumed by local residents. In Xishuangbanna, Yunnan Province, the average peasant household consumes 3.38 m^3 of timber a year. Timber used to cure tobacco amounts to 2 million m^3. One-third of annual net timber production is used as a source of fuel.

Demand for farm products has steadily risen with the increases in population. Grain production, in particular, has become a central focus of government policy: "Take grain as the key link." In the past, China was overzealous with the campaign, and other farming, forestry, animal husbandry, and fishery industries suffered neglect. Regardless of the topography, grain production became the all-important priority: "Produce grain in the mountains." Additionally, people in rural areas were asked to stop eating commodity grains. Large forested areas were either destroyed to produce grain or neglected, aggravating hydrologic cycles and soil erosion. On Hainan Island, several thousand hectares of forest were destroyed. Annual average losses of timber amounted to 200,000 m^3. In Dingqing, Luobai, and Suibin counties of the Three Rivers Plain, 25–67% of all forests were typically damaged from cultivation in the 1970s. In Guanxi Province, forests destroyed by reclaiming undisturbed land average 60,000 ha per year.

Transforming forested areas into cropland has damaged the environment. The Yellow River Valley epitomizes this degradation. The upper and middle reaches of the river valley were covered with verdant trees and plants 4,000 to 5,000 years ago. The landscape has since changed to barren hillsides, gullies, and endless hectares of loess soils. The river has since become dangerous. In the past 2,500 years, the river broke over its banks more than 1,500 times and experienced approximately 50 major changes in course. At present, the discharge of silt totals 1.6 billion tons with each cubic meter of the river containing 37.7 kg. As the quantity of silt deposit increases, the riverbeds rise at an average 10 cm per year in the lower reaches. At certain sections, riverbeds are 3 to 10 m above the ground. Floods consequently threaten vast areas of land, people, and businesses. Soils in the Loess Plateau have lost many of their nutrients and ability to conserve water. Agriculture and animal husbandry are therefore unstable and impoverish the region.

Forest degradation moreover threatens irrigation and other large state projects. Many projects have been put on hold for fear of siltation and other complications. The storage capacity of the Three Gorges Reservoir on the Yellow River, for example, has been 80% filled with silt. Furthermore, dams already built are in imminent danger. Large sums of money have been spent to dredge and alleviate silt buildup. Silt has blocked many of the small tributaries, thereby decreasing navigable river areas and contributing to elevated riverbeds and potential flooding.

Dwindling forested areas influence climate and especially rainfall in many areas. According to research in the Changbaishan Nature Reserve, the annual average rainfall has fallen by 6.8 cm in 20 years: it was 64.3 cm in the 1950s, 59.7 cm in the 1960s, and 57.5 cm in the 1970s. After deducting other factors, the decrease from deforestation amounts to 0.15 cm per year. Eastern areas of the region (near the Commonwealth of Independent States [C.I.S.], Japan, and Korea) have in fact experienced an increase in rainfall. Areas around Liaoning Province have undergone little to no change in rainfall. Only the central forested areas of the Changbaishan Nature Reserve have experienced declines in annual precipitation. This suggests that a major reason for the drop in precipitation is deforestation in the central areas.

Deforestation has also increased the frequency of natural calamities. In the nineteenth century, the Chaoyang District of Liaoning Province was well forested. At that time, there were 46 bumper harvests, 40 average harvests, and 14 suboptimal harvests. Deforestation drastically changed climate conditions and therefore crop yields. In the 23 years from 1958 to 1981, there have been 18 suboptimal harvests; moreover, 4 of the 6 years from 1980 to 1985 experienced drought conditions. The droughts continuously worsen in the North China Plain because of massive deforestation, as experienced in the Taishan Mountains. Successive years of drought have made the Baiyangding Lake, once the "pearl of North China," dry. Hundreds of thousands of fishermen now make a living by weaving mats. In the Southwest, deforestation has transformed once-favorable weather into drought and flood conditions. In recent years, mud rock flows, landslides, and flooding in Sichuan Province have occurred regularly. On the Yunan-Guizhou Plateau, as well, many areas have become bare with frequent flooding and drought.

Accounting for Forest Resources

As a renewable resource, forests are an invaluable natural asset. For a long time, the national forest policy has abused the asset; annual outputs of timber have been the main criteria with which to evaluate forest enterprises. This sort of accounting has endless limitations, among them underestimating residual forest values. The amount of investment allocated to harvesting trees has traditionally been greater than 70%, whereas the amount allocated to reforestation projects has typically exceeded only 20%. In spite of limited replanting, low seedling survival and poor management still cause slow regeneration. According to statistics from 1978 on 15 provinces, timber harvests amounted to 5.89 million ha. Approximately 1.97 million ha was artificially regenerated and 1.40 million ha was naturally regenerated. Regeneration, then, occurred on only 3.37 million ha, or 57% of the land. For the past 30 years, artificial afforestation has occurred on 90 million ha, of which only 28 million ha, or 30.6%, have been a success.

An important economic factor causing deforestation is the lack of reinvestment and government compensation for forested lands. The total value of forest reserves in China is RMB 1,292.99 billion yuan. Consumption on an annual basis is RMB 61.01 billion yuan, or 4.7% of the total value. Whereas state agencies and the timber industry may reinvest moneys for reforestation, consumption for fuel or deforestation by natural calamity do not command any money for reinvestment. Several funds, however, have been established to aid reforestation efforts. Although the funds increased twofold from RMB 310 million yuan in 1978 to RMB 680 million yuan in 1981, only 1.1% of the total amount was actually invested. A shortage of reinvestment money is a primary reason for continued deforestation in China.

Population densities in forested areas are high relative to world averages. A large population, poor management, and imperfect laws and regulations all spell disaster for the future of China's forests. Approximately 70% of the 17,000 forest fires from 1966 to 1979 were caused by human negligence. The fires destroyed 10 million ha of forests, a function of population density and poor public awareness. Derelict state foresters were also at fault for prescribing unauthorized fires. An exceptionally serious fire occurred in the Daxinganling Mountain Forest Region. The fire burned 1.01 million ha of which 70% was well stocked. The fire displaced 10,807 households and 56,092 people; it injured 226 and killed 193. Other casualties include 848,200 m^3 of timber, 2,488 pieces of equipment, 67 bridges measuring 1,340 m, 9.2 km of specialty railway, 483 km of communication lines, 284 km of transmission lines, and 3.25 million kg of grain. Intangible damages by smoke and dust are beyond estimation.

Forest pathogens, including disease and insects, are a kind of "fire without smoke." In 1980, forest areas attacked by moths totaled more than 3 million ha. The total loss in fiber was 4.3 million m^3, worth approximately RMB 200 million yuan. The frequency of forest disease and insect outbreaks is a function of tree stress and forest diversity. Diverse stands of trees without stress are better protected than monocultures experiencing stress.

Population Growth and the Distribution of Forests

Since 1949, the central government has frequently rallied the masses with slogans such as, "Plant trees to cover the country." China hopes to continue these mass movements now more than ever as a vehicle to reforest and improve China's environment. Artificial reforestation covers 14.51 million ha worldwide. China is among three countries with annual reforestation of more than 1 million ha (the other countries are the United States and the C.I.S.). Some of the more prominent reforestation efforts include "the five large shelter forest systems" and "the four great forest zones."

Three North Shelter Forest Systems

The North-Central, Northeast, and Northwest regions of China together cover 4.069 million km^2, or 42.4%, of the country's territory. The area, which includes 13 provinces and 551 counties, is weak in environmental protection and frequented by natural calamity. Desert generally exceeds arable areas. Soil erosion has impacted 554,000 km^2. Drought and periodic flooding occur frequently. Population growth rates exceed the national average by 40%.

The 3 North Shelter Forest systems, renowned as one of the greatest ecological projects in the world, were planned for these areas in 1978. The project has been divided into 3 distinct phases: 1978–2000, 2001–2020, and 2021–2059. Reforestation altogether totals 6.055 million ha; forest coverage has thus already increased from 5.1% to 7.1%. Protected farm and grasslands amount to 10 million ha. In Heilongjiang Province, increased forest coverage has been even more pronounced, rising from 4.8% to 13.0%. Natural calamities have gradually become less severe.

Reforesting parts of Beijing and Tianjin is an additional major goal of the project. Trees have already been planted in 76 counties around these two cities. The number of days with strong wind have fallen by 50% and sandstorms have become less of a problem. Soil erosion in the Loess Plateau should eventually be brought under control, as shortages in timber, fuel, feed, and fertilizer mitigate. The shelter forest systems will play a large role in rehabilitating the North's degraded environment.

Yangtze River Shelter Forest Systems

Two additional shelter forest systems line the Yangtze River, the third longest river in the world. Its drainage area covers 18.8% of the country and is home to 350 million people. The valley is one of the most inhabited and developed areas in China. Unsustainable harvesting and poor cultivation have degraded the forest cover in the middle and lower reaches of the river. The amount of soil lost annually to erosion is 2.24 billion tons. In a sense the river has evolved into another "yellow river," discolored by silt.

To alter its course of events, the government has established shelter forest systems in the upper and middle reaches of the river. Over the next 30 to 40 years, reforestation will impact 20 million ha. Tree distribution and age structure will integrate with socioeconomic and ecological characteristics of the region. The project encompasses 9 provinces and 145 counties. By the year 2000, forested areas will have doubled from 6.7 million ha to 13 million ha. Forest coverage will increase twofold also, from the current 19.9% to approximately 40%. Soil erosion should be brought under better control. Moreover, natural calamities of all sorts should become less likely.

Coastal Shelter Forests

A coastal shelter forest will also help to protect China's 18,000-km coastline. Eleven provinces and municipalities and 195 counties constitute a 25.1-million-ha coastal area. In spite of natural calamities that threaten industrial and agricultural production, the dense population and developed economy help make this region "the golden zone" of China. To lessen the threat of calamity, the government mandated in 1982 that shelter forests be encouraged. The project is to reforest areas with multiple tree species at different stories in the canopy. By the year 2000, forest cover should increase from the current 16.9% to 33.3%.

Taihang Reforestation

Another shelter forest project is taking place in the Taihang Mountain Range, which stretches from Shanxi to Hebei Province and includes the Wutaishan, Xiaowutaishan, Hengshan, Taiyueshan, and Zhongtiaoshan mountains. At present, forest coverage is low relative to that in the area previously. Natural calamities involving soil erosion and water shortages are a direct consequence of recent deforestation. The government has taken action by deciding on a plan to reforest the area by the year 2000. The project incorporates 3.60 million ha spread out over 110 counties. The aim is to increase forest cover from the current 15.3% to 43.6%. The project will also improve environmental conditions, agricultural production, and watershed conditions for Beijing and Tianjin.

Making the Plains Green

In the populated and developed plains areas of North and West China, arable land amounts to 50.6 million ha. Characterized by stable, high yields, the area covers approximately 38% of the country's major agricultural areas. Protecting this arable land is essential to stabilizing and further developing the region's economy. The government has called for and helped to implement a large-scale effort to "Make the plains green." For more than 10 years, efforts have contributed to the largest forest shelter for farmland in the world. Approximately 9.6 billion trees have been planted to green "the four sides" of fields. Intercropping has taken place on 4.4 million ha, and a total 13 million ha of arable land with an interconnecting network of forests has been successfully established. Forest cover increased from as low as 2% to 12.8%. By 1990, 363 of 918 counties already reached set standards at "making the plains green."

Forest Regions in China

China has 2 principal forested areas that are used both commercially and as shelter forests. The northeastern base includes Heilongjiang and Jilin provinces as well as parts of Liaoning Province and Inner Mongolia Autonomous Region. The central area of the forest encompasses the Daxingan and Xiaoxinganlin mountains, and the eastern areas of the Changbaishan and Wandashan mountains. The southwestern base includes the Yaluhe River Valley southeast of Tibet, western Sichuan Province, the Yunan-Guizhou Plateau, western Guizhou Province, and the upper reaches of the Taohe and Baihe rivers in the southern part of Gansu Province.

The northeastern forested area is the most productive in China. Timber yields exceed those in other areas by 30%. Most trees are cold-resistant conifers. In the Daxinganlin and Xiaoxinganlin forest zones, currently forested areas are more than 20 million ha in size, or 16% of the national totals. Total storage is approximately 2 billion m^3, or 21% that of the country. More than half of the commercial timber needed for construction is grown in these areas. Due to excessive logging, forests have suffered serious damage. If current trends continue, large-scale timber production in the Heilongjiang Forest Region will be forced to close within the next 10 years. To insure that this area remains a principal commercial and protection forest in the years to come, governmental forestry agencies are changing management strategies. They are beginning to spend more time on reforestation techniques and methods to preserve the fertility of cultivated areas.

The Southwest is the second major base for timber production. Large amounts of rainfall and rich diversity of species characterize this area. Various species of spruce produce 2,000 m^3 per ha. Because cutting historically has been relatively light in this area, the proportion of overmature trees is correspondingly large. The primary objectives in this area are to intensify silvicultural practices so that trees are harvested at sustainable levels while watersheds and soil quality are left largely intact.

To replant areas with fast-growing, high-quality trees is perhaps the single most important step that China can take to alleviate wood shortages and the environmental nightmares of large-scale unsustainable forestry. Intensive management of smaller areas is often more environmentally sound than partial cutting of larger areas. Presuming that the demand for wood is held constant, the former method might allow for larger areas to be set aside as wildlife preserves or undisturbed municipal watersheds. Forests south of the Dabieshan and Qinling mountains have characteristics conducive to such a management scheme. The region experiences tropical and subtropical climates. Chinese firs and bamboo should be the targeted species to attain the ultimate goal of a self-sufficient forest industry.

Intensive forestry mainly occurs in tropical and subtropical areas, as in the hilly areas south of the Yangtze River. Forest products from this area include teak, mahogany, camphor, rubber, palm oil, and coconuts. Fresh and dried fruits produced here are also important sources of nutrition for humans. Developing the fruit industry will enrich people's health. At present, per capita fruit consumption is 9.6 kg per year, less than one-sixth of the world's per capita consumption. Fruit industries that already have a foothold in China include oranges grown in the Yangtze River Valley, in Zhejiang, Jiangxi, Guangdong, and Guanxi provinces; and apples grown in Liaoning and Shandong. Pear, grape, Chinese chestnut, and walnut industries all have the potential to do well.

Coordinating Population Policies and Forest Resource Management

China's expanding population challenges initiatives to improve forest resource management. As the population grows and living standards improve, demand for wood products increases. Although trees are a renewable resource, the amount of land allotted to tree farms and natural forest ecosystems is finite. Moreover, conflicting short- and long-term interests spark debate on the best use for standing trees. Forestry in China confronts additional hurdles as follows: (1) competition with agriculture for arable land; (2) natural and artificial calamities, including insect infestations, diseases, fire, pollution, and anthropogenic damage; (3) insufficient investment; (4) poor management and disregard for contemporary science and technology; and (5) imperfections in the economy that involve lack of infrastructure, improper pricing, disregard for externalities, and similar shortcomings.

These and other major problems that confront China's forest management might be categorized as resource, ecological, and economic crises. Among the three, resource crises is preeminent. Its roots are in the mistakes of past management. For a long time, management plans viewed the science of forestry as simply production oriented. The role that forests play in soil retention and water conservation among other amenities was held in disregard. Overcutting and a lack of concern for regeneration have caused forested areas to dwindle in size. Environmental affairs as well as economic efficiency have accordingly deteriorated.

The notion that the sole end of forestry is to produce fiber for consumption needs to change. General environmental awareness of the many benefits of forest ecosystems may play a large role. Other efforts include the following: (1) fortify the status of the forest resource industry; (2) strengthen forest regeneration, conservation, and preservation; (3) mold management guidelines to local conditions; (4) improve pricing and valuation procedures; (5) establish a comprehensive accounting system; (6) compensate occupation and utilization; (7) continue research on ecological principles of forested ecosystems; and (8) improve public awareness of the many nontimber benefits of forest ecosystems.

At present, forest reserves in China are few. Excessive harvesting and insufficient regeneration have caused long-term deficits in stocking. Resolute measures need to be taken that protect and make rational use of existing forest resources. Ample human resources and cheap labor should work to the country's advantage. Mass campaigns might mobilize people to plant trees according to certain principles: plant species in concert with local soil and climate conditions, promote diversity, judge success by survivability, and reforest with the latest technology. China currently has more than 86 million ha of wasteland and hillsides suited for reforestation. If 80% of this area is reforested, national forest cover will reach 19.7%.

Low hilly areas in the 9 provinces and regions of South China enjoy abundant rainfall, mild temperatures, and plant and tree diversity. Trees grow fast, and the potential for plantation forestry is endless. Reforestation of barren slopes in Guangdong Province has enjoyed particular success. Other provinces and regions need to pay heed and follow suit. The establishment of 5 large protection forest systems, as previously described, is a fine start. These projects will certainly play a role in alleviating China's forest resource crises. Timber imports in the meantime should expand to fill shortages in supply for construction and other needs. Protecting the environment while pursuing China's "Four Modernizations" involves unremitting efforts and difficult decisions. The government should adopt the following four measures: establish fuel forests, improve administrative apparatuses, improve general stewardship, and prevent forest pathogens nationwide.

Fuel Forests

Nearly 900 million people live in China's countryside. A dearth in energy resources has caused many peasants to resort to straw and wood as primary sources of fuel. Developing efficient fuel forests is a necessary component of any management program. Managing stands of timber for this purpose alone will ease pressures on commercial and protection forests. It will further lessen the use of straw from crops for home cooking and eating. Fuel forests do not conflict with commercial and protection forests because they are more suited to places inappropriate to the latter, such as slopes and nutrient-deficient soils. Natural regeneration from stumps and fallen trees, or coppice regeneration, conserves soil and water. The social, economic, and environmental benefits of developing fuel forests are considerable. According to government estimates, China needs an additional 4.4 million ha of fuel forests to ensure an adequate fuel supply.

Forestry Administration

Administrative organs need streamlining and consolidation to effectively administer operations and orchestrate propaganda. In 1990, 72% of all townships and villages did not have administrative organs. All towns should have staffed stations to explain rules and regulations and to popularize standards. Weak enforcement and poor public awareness are the two primary limitations to otherwise effective legislation. More effective administration may help to limit forest degradation for the sake of fertile cropland. If existing cultivated areas are protected and productivity is improved, people might begin to make the following connection: "To reap grain one needs to control water, and to control water one needs to reforest lands." Soil erosion, wind erosion, and desertification make regeneration a challenge. In places where conditions allow, the population may be displaced to alleviate pressures on forest resources.

Improving Stewardship

Commercial, protection, and fuel forests should be identified in accordance with site conditions. Managers need to recognize where soils, hydrology, and climate establish a competitive advantage for one or the other type of forest. In broad terms, the Northeast and Inner Mongolia should focus on consumer, production-oriented forests. The Southwest and the southern portions of Gansu Province should specialize in establishing forests for soil and water conservation. Establishing consumer forests and tree plantations for timber production will also be a priority in this area. In Central China, "four-sided" protection forests and fuel forests should be a priority. In the North and northwestern parts of the country, protection forests need to control desertification and check winds, water loss, and soil erosion.

In addition to improving the distribution of forest types, harvesting methods need improvement. In principle, selective cutting is preferred to oversized clearcuts: "Remove the weak and keep the strong, remove the small and keep the large, remove the crooked and keep the straight, and give equal consideration to all." Annual harvests need to be limited to the region's annual growth. This will guarantee continuous, sustainable use and maintain the environment.

Lastly, forestry in China should gradually change from extensive to intensive management, and divorce itself from focusing only on timber production. China should make use of what some consider its competitive advantage—labor resources. Incorporating people in the country's forest management will not only benefit the industry and the environment, it will also help to alleviate China's "floating," unemployed population. Use of the nation's forest resources should be promoted while science and technology actively develop substitutes for wood to reduce certain types of consumption.

Preventing Forest Pathogens

Fire, insect, and disease outbreaks are serious threats to the nation's forest resources, especially as forest managers increasingly consider the benefits of plantation monocultures. Fire, disease, and insects will each destroy a stand of trees at quicker rates if the stand is composed of a single species. Forest fires in China are more commonly anthropogenic than natural. Most forest fires are caused by reckless use of fire; many outbreaks have been attributed to cigarette smoking and cooking. Insects and disease commonly spread if a stand is monoculture, under stress, or void of certain types of wildlife. Public awareness of the roles of wildlife and species diversity in preventing pathogens might help forests' self-defense.

7

Population and Grassland Resources

"Grassland" denotes meadows, grassy hillsides, and the steppe. Each of these landscapes is abundant in plant and wildlife diversity. Each plays a crucial role in the ecosystem's nutrient and energy cycles. Adaptation, rapid regeneration, and expansive ranges are defining characteristics of grasslands. They conserve water and soil, retard desertification, and prevent erosion by wind. Additional benefits and by-products include fodder, fuel, and industrial materials for economic development. Increasing demand for grain and decreasing availability of fertile soils are among the pressures to cultivate natural grassland areas. These pressures threaten unique grassland ecosystems of the desert steppe. Preservation and development need to be better coordinated to revitalize the national economy, consolidate national unity, and preserve the country's unique natural resources.

Grasslands are a renewable resource. They may be divided into three general categories: natural grasslands, improved grasslands, and artificial grasslands. Vegetation may be either annual or perennial. Grasslands help to sustain humans' livelihood by providing materials for animal husbandry, modern agriculture, and plant and wildlife diversity. They thereby improve general living standards. Grasslands also are a primary source for protein in the human diet. Herbivores such as sheep, oxen, and rabbits thrive off of grassy meadows and, in turn, provide nutrition to improve general human health. The per capita consumption of meat and milk is often a key indicator of living standards.

Aside from meat and milk, grasslands contribute to the production of leather, fur, and other livestock products. For nearly a century, developed countries have achieved major benefits by tapping their grassland resources. The 3.12 billion ha of grasslands worldwide constitute 20.9% of the world's land mass. As such an important resource, it is essential that the world protect grasslands from disappearing.

Over the ages, herdsmen from China's vast steppe have maintained self-sufficient life-styles. Once sparse, the area's population has grown in size. Herdsmen have since altered natural ecosystems to increase grassland productivity. In spite of the changes, people have learned to appreciate laws of sustained productivity and the multiple functions of this special ecological system. If humans disregard the carrying capacity of grassland ecosystems, overgrazing will lead to degradation and declining productivity. Temperature and moisture requirements are not flexible; the two factors are easily impacted by human intervention, affecting the stability of grassland ecosystems. Once impacted, grasslands may never regenerate. A single year or two of overuse or poor management may destroy vast areas of grassland. Grasslands may then require from 10 to 20 years to rehabilitate. To maintain an equilibrium among humans, animals, and grassland productivity, human and wildlife populations must be kept under control.

Grasslands are often considered "marginal" areas for cultivation. Agriculture and animal husbandry commonly border if not overlap grassland areas. Degradation is further aided by desertification and fire. Duststorms in the United States and sandstorms in China best exemplify disastrous consequences of grasslands degradation. In sum, efforts to preserve grasslands need to reconcile the benefits of preservation with the demands of China's current population growth rates.

Current Affairs

China contains 400 million ha of various types of grassland. This accounts for 40% of the country's total land base and 4 times the total area of cultivated land. Approximately 13.3% of the current cultivated land was initially grassland; this is the second highest national percentage in the world. In spite of the larger total figure, only 290 million ha of China's grasslands are arable. They may be categorized as follows: (1) grassy zones, 40 million ha; (2) drought zones, 53 million ha; (3) desert zones, 67 million ha; (4) the Qinghai-Tibet Plateau, 40 million ha; (5) the northern temperate hill zone, 24 million ha; (6) the subtropical, midland zone, 47 million ha; (7) the southern tropical hill zone, 13 million ha; and (8) shoal grassland zones, 13 million ha.

China leads the world in grassland diversity. This categorization scheme can be further broken into 18 groups, 37 subgroups, and more than 1,000 types of grass fields. Approximately 264 regions and counties in China practice animal husbandry. The variety of herbage in China also ranks first in the world. Approximately 1,130 leguminous and 1,150 xylophta grasses contribute to the more than 5,000 different species. Both the amount and variety of grasslands are, in fact, comparable to those in the United States. China, however, lags in management so that livestock by-products such as meat, milk, hides, and wool are far less than in the United States. Domesticated livestock raised in the United

States produces close to 9 billion kg of meat per year, accounting for 70% of the country's total. In comparison, China's livestock produces 0.45 billion kg of meat per year, accounting for only 10% of the country's total.

Grasslands play an important role in the national economy and people's livelihoods. Protection and rational use may alleviate crises from traditional stock raising, save supplies of grain, and enhance the land's carrying capacity. At present, pigs and poultry are the most common animals to raise domestically. They consume nearly one-quarter of total grain yields as fodder. A rising human population and decreasing amounts of cultivated land have contributed to instability in the current pig and poultry industries. The supply of wool for domestic consumption consistently falls short and necessitates imports of 200,000 tons at a cost of more than US$1 billion. Shortages in milk products exist as well.

If China were to take advantage of its grasslands as fodder, the country might increase its supply of meat, milk, wool, and fur. Additionally, China might conserve grain by feeding its pigs and poultry on legumes. Legumes are currently only 15–20% of the animals' total fodder. If these suggestions are implemented, the land's carrying capacity will improve, thus easing pressures on the environment. Rational use and protection will reduce deforestation, desertification, soil erosion, and inefficient water use. The development of agriculture, industry, mining, and tourism may also benefit.

An increasing population, decreasing amounts of arable land, and rising standards of living demand greater yield from grasslands. Conversion of grasslands into cultivated land and fields for animal husbandry has had a major impact, however, on China's total grassland resource base, especially on the steppe. Current trends suggest that these unique ecosystems are soon to be devastated.

China has historically manipulated its grasslands, especially in the North. Before the Qin and Han dynasties, minority nationalities scattered throughout the Northwest lived as nomads with little to no impact on their environment. Subsequent dynasties, however, adopted a policy of migration to reclaim marginal lands. This has increased population and accelerated environmental degradation. By the end of the Qing dynasty, stationed troops and desperate farmers headed north to reclaim land in the steppes. Grasslands suffered the largest destruction in history. Extensive areas were farmed and grazed as deserts encroached on once-vegetated areas.

Since 1949, improvements in medicine have helped to increase minority populations in pastoral areas. Arable areas have subsequently been cultivated to satisfy rising demands for food. Increases in industrial activity and mining also ensued. These developments have contributed to urban growth and further degraded grasslands.

Overly Populated Areas

Vast in size, rich in diversity, and relatively stable, grasslands are typically a battleground for competing interests, including forestry, farming, and animal husbandry. Since its inception, the Chinese government has encouraged migration to grassland areas. Moreover, the government has encouraged population growth of the minority populations. Unique grassland ecosystems are consequently heavily burdened.

As the largest contiguous steppe in China, Inner Mongolia has had a major population increase from 5.7 million in 1947 to 21.5 million in 1990. The growth rate is twice the national average. A closer look at the trends reveal four principal immigration peaks. In 1950, China attempted to jump-start its economy with the First Five-Year Plan. Within eight years, government projects contributed to an increase of 1.5 million people in the region. From 1958 to 1960, natural disasters elsewhere in China caused the second mass migration to the region. Approximately 2 million people traveled to the region in search of more suitable living conditions. China's infamous Cultural Revolution caused a third mass migration from 1969 to 1971. Inner Mongolia was a targeted area, where educated urban youth worked in the countryside. Construction regimes alone contributed 100,000 people within three years. The fourth mass migration occurred from 1973 to 1974, at which time the Hulunbeir Prefecture alone witnessed an inflow of 260,000 people. Such large-scale immigration to grassland areas has rarely been experienced in history.

A delicate balance exists between human population density and biomass productivity. By internationally accepted standards, a typical grassland area may support 5 persons per km^2. Populations in grassland areas currently far exceed these set standards. On average, Inner Mongolia is home to 15 persons per km^2. Even areas with poor productivity exceed the suggested standard, with population densities of 6.5 persons per km^2.

Grasslands Conversion

Slogans such as "No farming, no food" and "Win the fight between man and grass" characterize the prevailing mentality toward grasslands during China's Great Leap Forward and the Cultural Revolution. The lush steppe subsequently underwent drastic change. It is estimated that the total cultivated area of the steppe increased to 27 million ha, or 25% of the country's total cultivated area. One-third of this area is presently unproductive and void of vegetation. Another area, Heilongjiang Province, is one of the ten largest provinces with grasslands in China. It contains 6.7 million ha of grassland, accounting for 15% of the total province. Natural steppe accounts for 52.4% of the area's grasslands. In recent years, degraded areas have amounted to 646,500 ha, or 9.5% of the total grassland area.

Inner Mongolia contains 121 million ha of grassland. This is three-quarters of the region's area and one-quarter of the nation's total grassland area. Poor use and cultivation have seriously eroded grasslands in the region. During the Cultural Revolution, construction regimes cultivated 216,000 ha of grassland. Subsidiary food sources totaled 3.6 million ha, or one-tenth of the whole pastoral area. In Qinghai Province, 534 farms established from 1958 to 1960 encroached on grasslands. In Hebei Province, as well, an initial 800,000 ha of grassland was reduced by 84% to 130,000 ha because of excessive reclamation.

Animal Husbandry

Rapid population growth in grassland areas has tightened the food supply and impacted per capita incomes. People have had to increase their domesticated livestock to maintain living standards. The carrying capacity of grasslands limits grazing. Traditional animal husbandry, however, accounts for the number "in fence" and neglects the number "out of fence." Out-of-fence domestication lengthens animals' life span and leads to uncontrolled overgrazing and degradation. According to data from 10 pastoral provinces and regions in the North, the number of in-fence animals increased from 29.16 million in 1949 to 86.7 million in 1978. The average area for grazing per animal correspondingly decreased from 3.2 ha in 1949 to 1.3 ha in 1978.

The carrying capacity of Inner Mongolia is an estimated 69.4 million sheep. If utilization is presumed to be 60–70%, the amount of domesticated livestock falls to 48.4 million. The region's actual livestock population is as high as 70 million, surpassing the theoretical carrying capacity by 45%. In the early 1950s, the region averaged 0.25 sheep per ha. This grew to 0.8 sheep per ha after the mid-1960s and continued despite serious degradation. The increases in human population, then, have caused overgrazing, and overgrazing has reduced the region's carrying capacity. If continued, this trend poses a serious threat to China's grassland ecosystems.

Closer examination of one province may add insight to the complexities of grassland degradation. In the 1950s, Qinghai Province averaged 2.13 ha of pasture per sheep; this figure dropped to 1 ha per sheep by 1984, reflecting both the increases in domestic livestock and the decreases in pasture land. Output of edible grasses also fell in the 8 provinces surrounding Qinghai Lake from 25% to 43% within 15 years. Degraded, overgrazed pastures will on average decrease the body weight of oxen by 25–50 kg and sheep by 2.5–5.0 kg, and decrease wool by 1 kg.

Degradation of grasslands may be partly attributed to life-style changes of herdsmen on the steppe. In the past, China's grasslands were sparsely populated; herdsmen moved from place to place as nomads in search of water and grass. Pastures thereby had time to recuperate between settlements. Herdsmen and their families have since chosen to settle. Land is more commonly partitioned and individual households are responsible for maintaining the fertility of their overgrazed plots. Poor management and short-sighted planning have overburdened many areas. In northern Xinjiang, for example, inhabitants would travel with their livestock according to seasonal changes in climate. They would habitually travel to the desert steppe during winter and then move to Tianshan pasture and marshland areas in the summer. The cycle would protect the vegetation by leaving areas undisturbed for five to six months at a time. In Tianshan, townships and villages have increasingly divided land into pieces of private ownership. Regardless of the season, farmers maintain their stock inside the desert steppe and degrade pastures beyond the point of rehabilitation.

Population increases have exacerbated already serious fuel shortages in parts of China. Herdsmen chop fields, dig up root structures, and use stock dung as fuel. On average, a single household will clear 10 ha of grassland per year. According to this premise, approximately 5.3 million ha of grasslands are destroyed annually as fuel. Research in Wushan Banner, Inner Mongolia, shows that 700 households out of 13,700 use coal as fuel. The rest collect woody plants into faggots, burning 170,000 kg and damaging 33,000 ha of grassland each year.

Population growth and hunting have also seriously impacted food webs. Many birds of prey and other carnivorous animals have become nearly extinct. Rodents consequently multiply without predators. They gnaw at and dig up roots and soils, causing further grassland degradation. According to data from 1982, areas seriously impacted by rodents reached 67 million ha nationwide. Losses are estimated at more than RMB 1 billion yuan, or biomass equivalent to fodder for five million sheep over the course of a year.

Environmental degradation is an inevitable consequence of a growing human population and excessive grazing in pastures. Overgrazing will compact soils, reduce moisture levels, and induce desertification. Moreover, depleted vegetation decreases the soil layers' ability to absorb rain, resulting in soil erosion. Windswept sands then form dunes, cover additional fields, and possibly even blanket entire houses and villages. Strong winds and sand will also increase evaporation rates and contribute to aridity. The Chinese describe the scenario as "the desert marches on while human beings retreat." Research has shown that sandy areas of the steppe, caused by overgrazing, comprise 20% of the country's current desert area.

Poor Management

Aside from natural disasters, humans further contribute to grassland degradation with improper management guidelines, inaccurate scientific data, weak administrative institutions, and insufficient government investment in rehabilitation.

China has traditionally emphasized industry over agriculture, agriculture over animal husbandry, and animal husbandry over grassland preservation. These policies reflect a poor understanding of the importance of grassland resources in China's development scheme. A strategic management plan has yet to coordinate preservation and use of grasslands, which account for 40% of the country's total territory. Neglect toward the country's grassland resources occurs despite efforts to increase grain production, animal husbandry, and migration to western provinces. Moreover, the government has overemphasized a curative approach to national shortages in wool and milk by importing and accepting international relief; it has yet to focus on preventative measures involving investments to improve and protect the country's vast grassland resources.

Inaccurate scientific data cause poor understanding and erroneous hypotheses. For example, productivity of the land is commonly oversimplified and discussed in the media. Previously cultivated areas will generally not produce the same biomass as will sloped grasslands or poorly stocked woodlands. The actual productivity of areas will vary considerably according to topography, soil, hydrologic conditions, aspect, and management history. A Standardized Hectare Productivity of Cultivated Land (SHPC) helps to index actual productivity according to past agricultural use as follows:

Agricultural Use	Actual Productivity	SHPC Equivalent
Cultivated land	130 million ha	130 million SHPC
Grassland	220 million ha	11 million SHPC
Sloped grassland	67 million ha	4 million SHPC
Forest land	120 million ha	20 million SHPC
Sparse forest land	40 million ha	3 million SHPC
Freshwater bodies	5 million ha	2 million SHPC
Total	582 million ha	170 million SHPC

The amount of land used for agriculture, then, is equivalent to 170 million SHPC. Cultivated land accounts for three-quarters of this total. It follows that land, other than cultivated areas, harbors relatively little potential. In the North, for example, 20 ha of grasslands produce the equivalent of 1 ha of cultivated land. This low productivity is most likely a function of long-term exploitation and might be improved.

Hypotheses that claim grasslands insignificant are myopic. Among many other functions, grasslands support livestock for meat and other products. The country is characterized by small per capita areas of cultivated land and more vast areas of grasslands and mountains; China must therefore focus animal husbandry outside of agricultural areas. Investments in grassland conservation will realize large, although perhaps less visible, rates of return. Freeing additional areas of animal husbandry for cultivation and decreasing imports of animal products are two measures that will benefit the national economy.

In spite of the tremendous tasks for the management of China's grassland resources, agencies and personnel poorly coordinate interdepartmental planning and decisionmaking. Poor coordination has, needless to say, impacted progress in development, research, and training of appropriate personnel. At present, all sectors of natural resource management have established specialized administrative agencies at ministerial or departmental levels except for grasslands management. A Grasslands Division under the Animal Husbandry Bureau of the Ministry of Agriculture will not suffice. Grasslands cover 40% of the nation's territory and amount to 4 times the area of cultivated land. Under the present administrative framework, it will be impossible to reverse the trend of grassland degradation, enforce the grasslands law, and advance science and research.

Exploitation of grasslands will soon backfire without adequate government financing. Although grasslands produce animal products valued at RMB 3.9

billion yuan per year, the government invests less than RMB 0.1 billion yuan. The majority of this funding goes to disaster relief and capital construction. The moneys generally are distributed to small areas, neglecting large areas. To simply subsist, an increasing number of herdsmen continue to exploit the steppes resources. This myopia cannot last much longer.

Coordinating Population Policies and Grasslands Management

China has launched a nationwide campaign to preserve its grassland resources. Recent achievements provide encouragement to continue past efforts. An evolving, enlightened mentality as to the importance of grasslands has contributed to an administrative and legal framework for promoting preservation. Scientific research has further complemented and given substance to these efforts.

The change in thinking is perhaps the single most important turn of events. Scientists, technicians, and herdsmen alike are increasingly aware that grasslands degradation is primarily anthropogenic. An educated, concerned public will help to shape grasslands management. The renowned scientist, Qian Xuesen, has long been concerned with the protection and rational use of grasslands in China. Moreover, ideas concerning the development of grassland resources are presented in *On Resources*, a book by Li Jinchang recently published by the China Environmental Publishing House. The author proposes that grasslands management is of equal stature to agriculture and forestry; it thereby mandates bureau or ministerial status. Although the production output from grasslands is valued at only RMB a few billion yuan, the amount is forecast to climb to hundreds of billions in the near future.

Increasing public concern has already promulgated a national grasslands law and corresponding local regulations. The legislation protects herdsmen's rights and provides a legal framework for the protection and management of grasslands. Rural areas have additionally implemented a livestock "responsibility system." The contract system encourages households to practice animal husbandry in accordance with surrounding species of grass. Seed nurseries and testing stations have also been established to improve the quality and variety of seeds. Some varieties have in fact been sold on the international market.

Science, technology, and the training of personnel have also made progress in grasslands management. A map has recently been compiled that shows the geographic distribution of grassland varieties. The map lists forage varieties and presents monographs of the various types. Pilot projects, launched in 18 provinces, focus on grassland genetics, grassland ecology, remote sensing, zoned rotation grazing, and the impacts and processes of animal husbandry. Courses offered to the public are one of many different current activities to improve public awareness of the country's grassland resources.

China has actively pursued several measures to restore grasslands. Aerial seeding has impacted more than 1 million hectares in China; accompanying regulations have also been drafted to protect seeded areas. Since 1983, pastures for grazing have been established in several provinces to ease the impacts of grazing on natural grassland areas. The pastures have been established in Enshi (Hubei Province); Longshan, Yongxin, and Tongdao (Hunan Province); Liupanshui and Weining (Guizhou Province); and the northern grasslands of Inner Mongolia, Hebei, Qinghai, and Xinjiang. In recent years, eight protection zones and several grassland ecology stations have been established for protection and research.

Five measures have attempted to grapple with both the population problem and grasslands management. These include enforcing family planning in grassland areas, developing the potential for lucrative grasslands management, establishing artificial pastures for animal husbandry and agroforestry, improving species selection on artificial pastures, and establishing a responsible administrative network and adequate annual funding for protection initiatives.

Family Planning in Grassland Areas

In antiquity, grasslands were sparsely populated by minorities. Trends are rapidly changing. The average fertile woman gives birth four times. Although growth rates have rapidly increased, government authorities have exempted minorities from many of the country's family-planning initiatives. Population policies are therefore leniently enforced. Research has indicated that women should give birth to no more than two babies to minimize the impact of population on China's environment. Immigrant populations also need to be kept under control. Population growth will not cause socioeconomic and environmental problems in grassland areas if the growth is regulated and gradual.

Lucrative Grasslands Management

The public needs to become fully aware of the economic importance of grasslands protection. Its importance to national development is at least equal to that of agriculture or forestry. Science and technology should help farmers to choose the variety best adapted to a locality's soil, climate, and hydrology. General productivity will thus increase exponentially. The output of grasses in artificial pastures can be 4 to 6 times as large as in natural pastures. The protein component may also be 2 to 3 times higher. Aerial seeding of alfalfa, wheat grass, and other forage could be implemented. This might protect more natural, undisturbed areas and preserve ecological balances. Establishing artificial, highly nutritious pasture areas will encourage the production of high-quality wool and beef.

Protection zones and rotational stocking might help preserve desert grasslands. Eremophilous grasses such as Chinese tamarisk, desert date, and sand wormwood would also help establish high-quality breeds of goat, lamb, and camel. On the Qinghai-Tibet Plateau, efforts have been made to grow leguminous forage; establish artificial pastures; and breed goat, sheep, and yak. The measures have alleviated both environmental degradation and consumer demand. Water and soil conservation in other northern temperate zones are a primary concern. Artificial pastures grow alfalfa, ormosia, shrub lespedeza, and wild barley. These crops will in turn help breed cattle, sheep, and goat. China's hilly, subtropical areas are characterized by high productivity, warm climates, and ample rainfall. Mowed grasslands produce white clover which, in addition to rye grass, promotes cattle and sheep farming. The South's tropical climates produce elephant grass, India paspalum, and column vetch—grasses fit for cattle, sheep, and most other herbivorous domestic livestock. Coastal and riparian grasses include sesbania, alfalfa, wild rye, and fescue. These grasses are conducive to sheep and rabbit farming as well as grass powders.

Artificial Pastures

Natural resource management in China needs to have a broad scope and diverse objectives. For example, animal husbandry initiatives need to cooperate with agriculture and forestry departments to find solutions to environmental degradation. Relying on individual sectors and ignoring the crucial roles of other sectors is myopic and a recipe for economic and environmental disaster. Limiting reclamation for farmland is a good first step toward grasslands protection. Next, tree farms and pastures might be encouraged on lands exhausted by farming. Agroforestry, or "vertical" agriculture, is based on intercropping and should also be encouraged to maintain soil nutrient cycling. Artificial grasslands will both benefit animal husbandry and help maintain ecosystems. These combined efforts should help to establish a "green ocean" zone across the Gobi Desert.

Species Selection

New methods and technology need to be adopted to monitor changes in type, structure, productivity, and sustainability of grasslands. Research should concentrate specifically on species selection and determining levels of sustainable use. "Strong in summer, fat in autumn, slim in winter, and dead in spring"—the cycle paints a bleak picture for species survival. Combining traditional animal husbandry with intensive farming and winter pastures might revert the trends. The scientific selection of which species to plant will not only minimize shortages in forage grasses, but also will protect soils from overexposure during spring and winter winds. Shelter feeding might be justified in some areas to ease pressures on pastures.

Government at all levels should promote the use of solar and wind energies, which are ample in grassland areas. If herdsmen were able to harness these energies they would be less apt to mow and uproot grasses in the field for fuel. Widespread medicinal herb gathering is also damaging to grasses in natural pastures. Gathering should be strictly regulated or confined to certain areas on pastures already badly damaged.

Administration and Financing

The government needs to establish authoritative agencies to create, implement, and enforce innovative solutions to protect and sustainably use grasslands. Grasslands protection should additionally be incorporated in the annual planning for national development. A fund for grasslands management would help to finance efforts by farmers and herdsmen to responsibly use and protect grassland pastures. Projects might include fencing, irrigation, seeding, rodent extermination, and natural disaster prevention.

Summary

In sum, grassland management, above all else, needs a strategic plan that incorporates growth in the human population, agriculture, animal husbandry, urban construction, and forestry. The "three combinations" presents the following preliminary framework: (1) artificial pastures and animal husbandry with processing; (2) scientific research and training with production; and (3) animal husbandry and industry with commerce. Reforms in public awareness, farming methods, science and technology, administration, and financing will together contribute to grasslands rehabilitation and protection.

8

Population and Mineral Resources

Ever since tools were carved from stone, civilization has relied on the use of mineral resources for survival. The Industrial Revolution, in particular, marks a sharp increase in mineral resource extraction and use. Resource extraction has since accompanied rapid population growth and economic development. Industrial growth has consumed considerable quantities of nonrenewable resources. The extraction has altered energy cycles in that humans have come to depend on excess energy from ages past. Rates of energy consumption now far exceed rates of energy deposition; in other words, more energy is required, or used, than is currently created. The buildup of greenhouse gases, ozone depletion, and acid rain each attest to this shift in equilibrium. Large-scale consumption of mineral resources is a preeminent environmental concern.

Mineral resources include substances in their solid, liquid, or gaseous state that are formed naturally and are of value to humans. They include not only those deposits already identified but those that potentially exist. The term refers to exploitable resources, both current and future. Mining potential reserves is, of course, a function of consumer demand and technology. In spite of this definition, national economic audits typically account for only proven reserves.

In retrospect, mineral resource extraction has grown hand in hand with the population. Humans currently tap energy reserves from the past to accommodate a ballooning population. Ecologists estimate that the earth's carrying capacity amounted to no more than 10 million people during the Stone Age, when population densities hovered at 0.01 to 1 person per km^2. The advent of agriculture and use of iron and copper tools increased this carrying capacity by a factor of 500. Population densities subsequently reached as high as 25 persons per km^2.

Since the Industrial Revolution, large-scale use of fossil fuel energy has had a pronounced effect on the environment. Increased productivity has expanded the earth's carrying capacity for humans. Mineral resource industries have become a cornerstone for human life and productivity. In China, 90% of the energy and 75–80% of industrial raw materials come from mineral resources. Figures in industrialized, developed countries even exceed these. Sharp population growth over the last few decades might not have occurred had humans not tapped the earth's excess energy reserves (see Figures 8.1 through 8.3).

In antiquity, small-scale mineral resource extraction barely influenced demography. The Industrial Revolution, however, and the use of coal and iron induced major changes in world demography. Cities took root in areas rich in coal, iron, and other mineral resources. In the United Kingdom, the United States, France, Germany, and the former Soviet Union, industrial complexes have influenced the demography of past, present, and future generations. Ruhr, Germany, is abundant in coal and iron; it exemplifies the impact that mineral resource deposits have on demography. One of the largest uncovered coal deposits in the world, the mine in the Ruhr Valley has operated for a century and is expected to remain active for another three. Throughout the early 1980s, the population density in the Ruhr region ranged from 1,300 persons per km^2 to 3,000 persons per km^2 in more central areas. Within the compound, small and large cities are intertwined. Approximately 19 cities have populations in excess of 100,000; 4 in excess of 500,000; and the largest has a population of about 1 million people. In China, the city of Anshan in Liaoning Province is considered the nation's steel capital, and Fushun is considered the coal capital. Benxi, Tangshan, and Manshan are additional important iron-, steel-, and coal-producing centers.

Figure 8.1 Consumption of Iron Ores in Relation to Population Growth

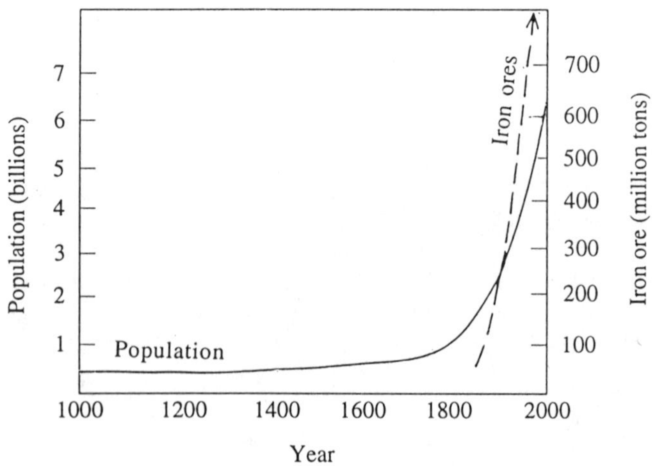

Source: *Ecological Research*, 1984, Gansu People's Publishing House.

Figure 8.2 Consumption of Lead Ores in Relation to Population Growth

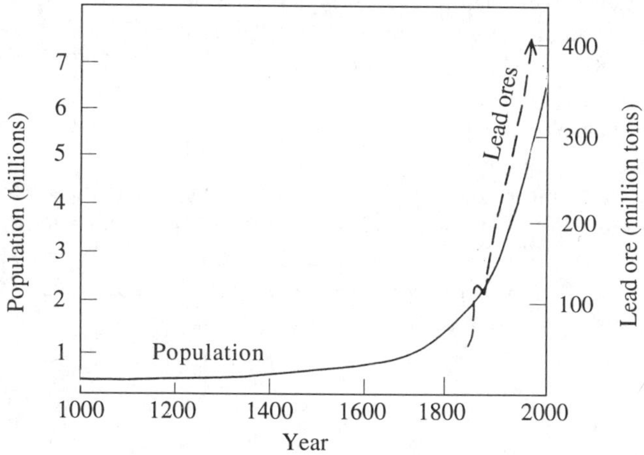

Source: *Ecological Research*, 1984, Gansu People's Publishing House.

Figure 8.3 Consumption of Copper Ores in Relation to Population Growth

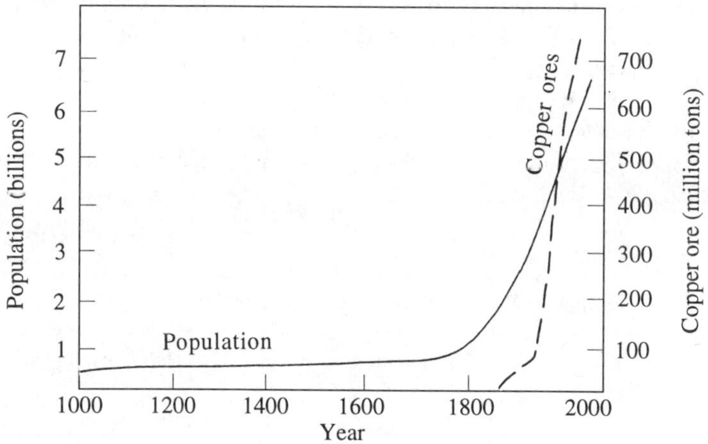

Source: *Ecological Research*, 1984, Gansu People's Publishing House.

The discovery and use of mineral resources is vital to the growth of the world economy. It also, however, contributes to environmental degradation, distorts natural energy flows, and exposes toxins. Strip mining changes surface structures by peeling off layers of rock and soil, exposing soils, and disturbing groundwater flows. The residual occupies large areas and pollutes the air and water. Ore cleaning, smelting, and processing introduce large amounts of toxic gas, contaminated water, and dust to the environment. Mining for mineral resources continues at unprecedented levels: 1.22 million tons of coal per day (1987), 8.18 million tons of crude oil per day (1980), and 1.4 million tons of iron ore per day (1980). Needless to say, mining at these levels has had a tremendous impact on China's natural surroundings, as well as on its remaining resource stocks.

The data for iron, lead, and copper mining suggest that increasing per capita consumption will cause resource exhaustion at sooner than expected dates. Table 8.1 shows the results of research on supply and demand dynamics that determined the years of peak production and the years of resource exhaustion based on current knowledge of mineral deposits.

Current Affairs

Once rich in mineral resources, China has fallen into an increasingly serious mineral resource crisis with its ever-expanding population. Per capita reserves are in fact small relative to the world average—China ranks 80th in the world. Only 148 sites out of a total of 14,000 are considered rich in reserves. The country leads the world, however, in its variety of mineral reserves. Approximately 162 minerals may be extracted from 200,000 different sources. Supply and demand statistics for most minerals are less than optimal. Low-consumption minerals are rich in reserves whereas other staple minerals are lacking. Abundant minerals include tungsten, molybdenum, antimony, tin, coal, graphite, magnetite, talcum, and other rare earth metals. Less abundant reserves include iron, manganese, lead, copper, aluminum, zinc, nickel, gold, silver, sulfur, phosphorus, uranium,

Table 8.1 Estimated Peak Production and Exhaustion for Minerals

	Peak output year	Exhaustion year
Aluminum	2060	2215
Chrome	2150	2325
Gold	1980	2075
Lead	2030	2165
Tin	2020	2100
Zinc	2065	2250
Asbestos	2015	2150
Coal	2150	2405
Crude oil	2005	2075

Source: *Journal of Ecology*, Vol. 2, 1983.

and crude oil. Scarce reserves include chromium, cobalt, platinum, sylvite, boron, and diamond. Only coal is consumed in large quantity.

Staple minerals in China are commonly lean ores. Although rich ores have been discovered, they do not include minerals such as iron, manganese, aluminum, copper, gold, sulfur, phosphorus, and uranium, which are vital to economic construction. For example, 86% of iron deposits are lean ores with grades of 30–35%; only 35% of copper deposits have grades over 1%, and 6% have grades over 2%; finally, only 6.8% of phosphate deposits are rich, containing more than 30% of P_2O_5. China has far more associated deposits than it does single, contiguous mines. One-fourth of iron mines, two-fifths of gold mines, three-fourths of silver mines, and four-fifths of nonferrous metal mines are associated deposits. China generally has more small to medium-sized mineral deposits than it does large ones. Large deposits account for only 8.6% of all mines.

Uneven distribution is an additional characteristic of China's mineral deposits. Coal is primarily found along a belt running from the northern provinces of Shanxi and Shaanxi into Xinjiang and Inner Mongolia. Phosphorus deposits are concentrated in Yunnan, Guizhou, Sichuan, and Hubei. Lastly, iron is found in Liaoning, Hebei, Shanxi, and Sichuan.

Problems

China has long exploited its coal, iron, copper, tungsten, tin, gold, silver, lead, zinc, antimony, mercury, and arsenic deposits. Per capita consumption prior to 1949 was in fact minuscule. Since that time, however, more than 100 minerals have been discovered and used in one capacity or another. From 1950 to 1987, the total output of minerals increased by a factor of 36, from 50 million to 1.8 billion tons. Outputs of crude oil, coal, iron ore, nonferrous metals, tungsten, and phosphorus grew at annual rates of 8–19% (see Table 8.2).

China has become an increasingly prominent producer and processor of minerals (see Table 8.3). Resource consumption has also grown, however, at faster rates than growth of national income. From 1953 to 1985, national income grew by a factor of 6.25, whereas resource consumption grew by a factor of 13.23. Although per capita consumption of resources remains deceptively low, aggregate consumption is in fact quite large. A major reason for China's massive consumption lies in its industrial inefficiencies. The amount of resources needed to increase gross national product (GNP) far exceeds the amounts needed in more developed, industrialized countries. The ratio of energy consumption to GNP in China is 10–38% that in industrialized countries; energy intensity, however, is 3–10 times that in other countries. Potentially large savings might be made if extravagant consumption can be controlled.

Given China's industrial expansion and current stage of development, improved efficiency will not reverse trends in mineral resource consumption. Moreover, studies in other developing countries indicate that increasing demands for raw materials usually outpace GNP growth when per capita GNP is

Table 8.2 Mineral and Mineral Product Production, 1950 and 1985–1987

Product	1950	1985	1986	1987	Average growth rate, 1950–1987 (%)
Steel (10^4 tons)	61	4,679	5,220	5,628	13.0
Pig iron (10^4 tons)	98	4,383	5,068	5,503	11.5
Raw coal (10^4 tons)	4,300	87,228	89,400	92,800	8.7
Crude oil (10^4 tons)	20	12,489.9	13,061	13,414	19.6
Iron ore (10^4 tons)	235	13,819	11,950	16,143	12.2
Nonferrous metals (10^4 tons)	3.06	155.8	167.3	196.6	11.9
Copper	1.05	41.3	47.6	51.64	
Aluminum	0.91	52.5	55.9	61.62	
Lead	0.49	223	23.0	24.64	
Zinc	0.07	30.6	31.9	38.31	
Antimony	0.59	3.57	3.44	6.68	
Tin	0.86	2.84	2.91	3.02	
Wolfram ore (85%) WO_3	1.19	4.88	5.03	5.33	
Sulphur iron ore (10^4 tons standard)	10	691.9	783.5	1,054.9	12.8
Phosphate ore (10^4 tons standard)	2	680.7	979.8	1,486.6	18.8
Chemical fertilizer (10^4 tons)	1.5	1,322.2	1,395.7	1,672.2	21
Cement (10^4 tons)	1.41	14,594.8	16,606	18,625	14.1

Source: *Theory on Resource Auditing*, Maritime Publishing House.

Table 8.3 China's Primary Minerals

Products	World total (10^4 tons)	Output of China (10^4 tons)	China's output as percentage of world output	Rank	Three largest producers
Steel	65,500	4,679	7.1	4	USSR, Japan, U.S.
Coal	433,600	87,228	20.0	2	U.S., USSR
Oil	266,900	124,900	4.7	5	USSR, U.S., Saudi Arabia
Iron ore	90,510	13,819	15.3	2	USSR, Brazil
Copper ore[a]	830.1	26.5	3.2	9	Colombia, U.S., USSR
Bauxite[a]	8,830	140	1.6	11	Australia, Guinea, Jamaica
Aluminum ore[a]	356.2	21.5	5.9	5	USSR, Australia, U.S.
Zinc ore[a]	691.8	37.1	5.4	5	Canada, USSR, Australia
Wolfram ore	5.38	2.32	43.0	1	USSR, Canada
Tin ore	19.68	2.98	15.0	2	Malaysia, Brazil
Silver ore	13,321[b]	520[b]	3.9	15	Mexico, Peru, USSR
Gold ore	1,528[b]	48[b]	3.1	7	South Africa, USSR, Canada
Manganese ore	2,274.2	160	7.0	5	USSR, South Africa, Brazil

a. Metal content.
b. In tons.

below US$2,000. According to other sources, when industrial output commands the larger part of a country's GNP, demand for resources grows at higher rates. In most developing countries, a 1% increase in GNP typically necessitates a 2% increase in steel consumption, a 1.7% increase in consumption of other metals, and a 1.1% increase in energy consumption. In light of industrialization, China will be pressed to slow growth rates of mineral consumption. Likewise, it has very little influence on demand outside of its borders. In the next century, China will pass through its periods of peak resource consumption. Industrial growth and population increases will certainly exert strong pressures on mineral resource stocks as overconsumption becomes a primary environmental concern.

In spite of China's early stage of economic development and its low per capita consumption, the country has already become a major producer and consumer of mineral resources. The present annual production of mineral ores amounts to 5 billion tons. Total output of minerals has surpassed levels in the United States. Overburdened with a burgeoning population and still at an early stage in the development process, China will have a far more difficult time reducing mineral resource consumption than will developed countries.

Pressing demand for mineral resources has not only induced shortages, it has also caused wanton ecological destruction. A large population demands minerals in their processed state. The demands have led to large-scale mining and raw material processing that discharge significant amounts of contaminated water, gas, and air (see Tables 8.4 and 8.5). Polluting industries are prematurely threatening China's path to development with environmental catastrophe. Western industrialized countries, by comparison, encountered problems with their environment in the 1960s and 1970s, by which time their national income had already ballooned. China is therefore at a relative disadvantage in having to pay for environmental protection at an earlier stage in the development process.

Trends

Development trends indicate that per capita consumption of resources will grow in China. Pressure to extract increasing amounts of mineral resources augur ill

Table 8.4 Emissions from Mining and Processing, 1985

Industries	Gas ($10^8 M^3$)	Particles (10^4 tons)	SO_2 (10^4 tons)	NO_x (10^4 tons)
Total	62,746.6	6,013.5	1,252.9	491
Iron and steel	5,192.9	243.8	122.2	48.1
Nonferrous	891.8	37.0	19.0	7.5
Building materials	6,345.8	279.0	141.4	55.7

Source: *Environmental Studies*, Volume 5, 1989.

Table 8.5 Emissions from Industrial Production, 1985

Industries	Industrial waste gas ($10^8 M^3$)	Dust (10^4 tons)	SO_2 (10^4 tons)	NO_x (10^4 tons)
Total	16,960.7	1,439.14	101.11	1.21
Iron and steel	6,433.79	452.73	8.76	—
Nonferrous	1,740.19	185.56	53.96	0.87
Building materials	—	798.68	—	—

Source: *Environmental Studies*, Volume 5, 1989.

for the environment. As of 1990, China had a population of more than 1.1 billion. This figure will exceed 1.25 billion by the end of the twentieth century. To match current per capita consumption of resources in the United States, China would have to quadruple its mineral output. A U.S. magazine, *Earth Science*, reported that the annual per capita consumption of minerals, excluding reusable metals and building materials, is 20 tons. Along these lines, the total annual consumption of minerals in China will exceed 25 billion tons—more than current consumption in the United States, China, and the C.I.S. combined. These levels are unfeasible and would inflict disaster to the environment.

China needs to refocus its efforts so that progress continues at steady, low per capita mineral resource consumption. The ratio of resource consumption to GNP suggests that present measures will have a pronounced effect on China's development and on its environmental affairs. If the mining and processing of mineral resources becomes more efficient, China will have taken a significant first step.

Increases in mineral consumption worldwide are more related to per capita consumption than to population growth. For example, after World War II, the annual growth rate of mineral production and consumption averaged 5%, far higher than the average rates of population growth. Whereas population growth typically occurs in developing countries, consumption growth typically occurs in developed countries. A widening gap thus has emerged with regard to per capita consumption of minerals in developed and developing countries (see Table 8.6).

The oil crisis of the 1970s prompted developed countries to curb consumption of their mineral resources. Industrial expansion in developing countries did not allow as much flexibility in response. Population growth in these countries only compounded their plight. China in essence epitomizes developing nations who have since the 1970s found themselves trapped by population growth, recalcitrant industrial expansion, and natural resource crises. Between 1980 and 1988, approximately 7.8 million tons of steel, 200,000 tons of nonferrous metals, and 21 million tons of cement were consumed in China to support the more than 100 million people born during this period. Into the twenty-first century, China

Table 8.6 Per Capita Consumption of Major Minerals, 1966–1969

Country	Population (millions)	Crude steel (kg/person)	Refined copper (kg/person)	Aluminum (kg/person)	Zinc (kg/person)	Gross energy (coal equivalent)
United States	203	660	9.3	16.9	6.0	10,000
Other developed countries[a]	285	366	4.9	7.3	3.3	3,500
Developing countries	2,445	18	0.15	0.2	0.17	330

a. European countries, former Soviet Union, Japan, Canada, Australia, New Zealand, Israel, South Africa.

Source: G. M. Master, *Introduction to Environmental Science & Technology*, Science Press, 1982.

will have to further expand the scale of mining processing and importation to meet the needs of a growing population.

At present, the population surplus in rural areas poses serious problems. Burgeoning rural populations tend to overuse cultivated areas, forests, grasslands, and mineral deposits. Western areas in particular are already in short supply of arable land. An excessive population causes surplus labor and extreme population densities in areas with mineral deposits and processing facilities. In Luanchuan County, Henan Province, for example, people subsist on 0.052 ha per capita. Government subsidies and cheap loans for agriculture are the only vehicles to keep the area afloat. Annual per capita income is approximately RMB 230 yuan. One of the three largest mines in the world—the Luanchuan molybdenum, lead, and zinc mine—attracted a deluge of miners. Within several years, miners established 988 shafts and 1,934 quarries. A single open cast lot of 0.2 km^2 in size requires 103 enterprises to produce 5,000 tons of ore. Approximately 200,000 of the 240,000 inhabitants of the area are engaged in mining.

Unauthorized mining has caused grave waste and inefficiency. Toxic discharge from these operations pollutes four principal rivers in the county. In Yunping, Yunnan Province, proven reserves of lead and zinc total 14 million tons. An estimated 10,000 miners have rushed in to dig out high-quality ores. Approximately 330,000 tons of ore have been mined to date at an expense of 5 million tons of reserve—a less than 7% recovery ratio. Unauthorized mines are scattered throughout China. Unless this surplus, unauthorized labor is steered elsewhere, the problems of waste, inefficiency, and increased pollution will persist.

Shortages

Problems stemming from China's huge population are matched by shortages in mineral resources. Shortages have recently stretched the country's energy supply and quickened the exhaustion of mines. Mines are rapidly depleted and recovery/utilization ratios are poor.

Unauthorized mining rapidly depletes otherwise well-stocked mines. Individual miners, townships, and whole villages will flock to discoveries of ore and cut the longevity of mines. At Luanchuan, for instance, a mine life once estimated at 100 years may soon be reduced to 10 years if mining activities continue as at present.

Moreover, the recovery ratio of minerals is extremely low. Large quantities of minerals are wasted as residual. According to annual reports on mining, the recovery ratios are 50–60% for state-owned mines, 30–40% for local mines, and 10–20% for coal pits. If the average recovery ratio is 40%, then the utilization rates of coal decline to 10–20%. The average heating efficiency of boilers, furthermore, is 55%. The utilization ratios are far below those in other areas of the world, such as Japan (50%), the United States (50%), and the countries in Europe (40%).

Presuming recovery rates of 65–69%, the overall utilization ratio of iron ores is only 36.7%. Presuming a 50–60% recovery rate in mining, selecting, and forging, the utilization ratio for nonferrous metals is 25%. The recovery ratios for nonmetals range from 20–60%. Utilization ratios of China's mineral resources in general are 10–20%, lower than those in industrialized countries.

Such low rates cause the loss of associated deposits. Among 1,845 developed mines, only 2% exceed a 70% utilization ratio; less than 15% reach 5% utilization; and 70% stay below 2.5% utilization. Overall recovery for mines that contain copper, lead, zinc, and nickel range from 10–30%. Developed countries such as the United States and Japan boast more efficient overall recoveries of from 66% to 90%.

Recycling efforts are minuscule and present a real opportunity to conserve and improve efficiency. From 1981 to 1985, recycled copper, lead, and aluminum scraps amounted to 51%, 32%, and 44%, respectively, of total consumption in Western countries. Outputs of steel produced from scraps have risen to 50% of the total. Recycling in China, however, has yet to take root as it has in some European countries. During the Sixth Five-Year Plan, recovered steel scraps totaled 103 million tons; approximately 20% of this total was actually recycled and finally accounted for 8% of the total steel consumed. Recycled nonferrous metals accounted for only 11% of the total output; this is lower than world average by a factor of 3 to 4. Mineral resources are typically dumped into natural surroundings as contaminated gas, water, or trash. This is the preeminent cause for intense pollution at China's low stage of development.

Coordinating Population Policies and Mineral Extraction

To ease tensions between a huge population and an insufficient mineral resource base, China has made substantial efforts at increasing supply and decreasing demand. In 1949, China surveyed mineral reserves to ensure industrial expansion; nearly 20,000 ore zones were discovered. Over the past decade, management and protection of these mineral resources has been mandated by legislation. The Chinese government also has laid out guidelines for establishing individual, township, and village-operated mills. These steps have reduced the number of unauthorized, reckless mining operations that degrade the environment.

According to specialists, large mineral resource deposits remain uncovered. It is crucial that China discover such resources and mine them at more sustainable rates. At present, approximately 10 of 45 important minerals will be in short supply by 2000. Among 15 major minerals—coal, crude oil, natural gas, uranium, iron, bauxite, copper, lead, zinc, gold, sulfur, phosphorus, potassium, sodium, and materials for cement—6 are either low in identified reserves or are currently inaccessible. Chromium, cobalt, platinum, silver, manganese, boron, and diamond are also in short supply for China's Eighth and Ninth Five-Year Development Plans. The concern is that even international markets will not be able to provide a sufficient amount of the necessary resources. Not only are China's importing capabilities limited, then, but the country is reliant on a market supply. Constricting global supplies further hamper this possibility.

The ultimate object is to continue development and economic expansion while maximizing environmental protection and sustainable, efficient use of resources. Five measures might help achieve the objective: (1) continue economic development to finance protection, (2) increase public awareness of mining legislation, (3) examine and define property rights issues, (4) reassess pricing, and (5) improve resource auditing and accounting.

Economic Development

China's enormous population makes its path to industrialization unique. Industrial and economic development depends on highly intensive resource consumption. Only in later stages of development can China shift gears to focus on reducing waste and broadening its resource base. At this stage, it will work to ease its mineral resource shortages and ensure a more balanced supply and demand of raw materials. Broadening the supply of mineral resources may involve the following: (1) increase funding for exploration, (2) develop mines at all scales, (3) enrich lean ores, (4) improve mining efficiency by using associated mineral deposits, (5) maximize specialization and free trade by using both foreign and domestic suppliers so that imports and exports are balanced, and (6) develop substitutes for rare minerals with science and technology.

Cutting waste is as important as broadening the country's resource base. It follows that laws and regulations to protect mineral resources need to focus on the mining and processing procedures in addition to focusing on undisturbed areas rich in resources. Reducing consumer demand is also as important as regulating the supply. Along these lines, science and technology can promote efficiency by reducing industry's raw material consumption. The overall aim should be to raise utilization levels and reduce waste. China is left with few alternatives but to make full use of its resources in markets at home and abroad. Government might encourage conservation of resources as well as employ the country's floating population by readjusting priorities in general administration, industry, research and technology, and foreign trade.

Public Awareness

Massive campaigns should help to educate people about the population and natural resource crises that surround them. Radio, television, newspapers, magazines, and school textbooks can convey the message that China's per capita resource stocks are low in spite of large aggregate stocks. Moreover, the public needs to be aware that many of China's resources are not renewable, but finite. Once public concern is generated, people will be more willing to make sacrifices to conserve resources and recycle when possible.

Property Rights

Property rights need to be more clearly defined. Although the Constitution and China's Mineral Resources Law stipulate that all mineral resources are owned by the state, this principle has never been fully appreciated. Many of China's mineral resource deposits are subject to unlawful possession because of unclear property rights and the randomly distributed nature of mineral deposits in the first place. Unconcerned with ownership titles, miners care more about making quick money than the long-term interests of the country. This mentality quickens exhaustion of mines.

Government should strengthen its function as a supervising body by separating this function from actual ownership and management. If administration is detached from operations, property rights issues will become more clear. Any and all mining should thereafter require permits. Legislation and law enforcement should also be strengthened to encompass issues in exploration, exploitation, and protection. The rights and duties of mineral resource ownership need to be more clearly defined and ensured.

Pricing

China's current pricing scheme has strained domestic supply and contributed to industrial inefficiencies and waste. "High-priced products, low-priced raw materials, and priceless resources," characterizes the pricing structure of China's mineral resources. Costs of ownership, maintenance, exploration, and pollution are typically not included in the prices. Three measures might help establish just pricing: (1) revise general pricing theories and methods, (2) use more sophisticated valuation to establish mineral resource prices, (3) use pilot projects to draft and implement reforms. Any and all reforms should make owners accountable for resource rents or the marginal user cost of mineral deposits. If mineral deposits become more valuable as they exist in their natural state, rates of extraction will slow and resource deposits will last longer into the future. This sort of pricing mechanism will not make up for mistakes made in the past, but it will help to preserve what remains.

Accounting

Since mineral resources are the building blocks for national wealth, they need to be properly accounted for. Each generation should consume resources to survive and regenerate resources for future generations. This responsibility can be met only if people are aware of what resources they manage and at what rates these resources are growing or shrinking. Accounting for natural resources is an integral part of determining a nation's wealth. Audits are, additionally, a key indicator of socioeconomic progress and development. Three measures might help to establish a just accounting system: (1) current indices of mineral resource reserves should be modified to indicate readily available resources versus currently inaccessible resources, (2) accounts that value mineral resource stocks need to be established, and (3) the government representative that administers the mine in question should be paid compensation according to the mine's exhaustion. In other words, a depletion rate needs to be established that is similar to the depreciation of fixed assets.

Gao Guangde

Nature Conservatory, Jiuzhaigou

Gao Guangde

Yangtze River

Gao Guangde

Yellow River

Gao Guangde

The Utilization of Land

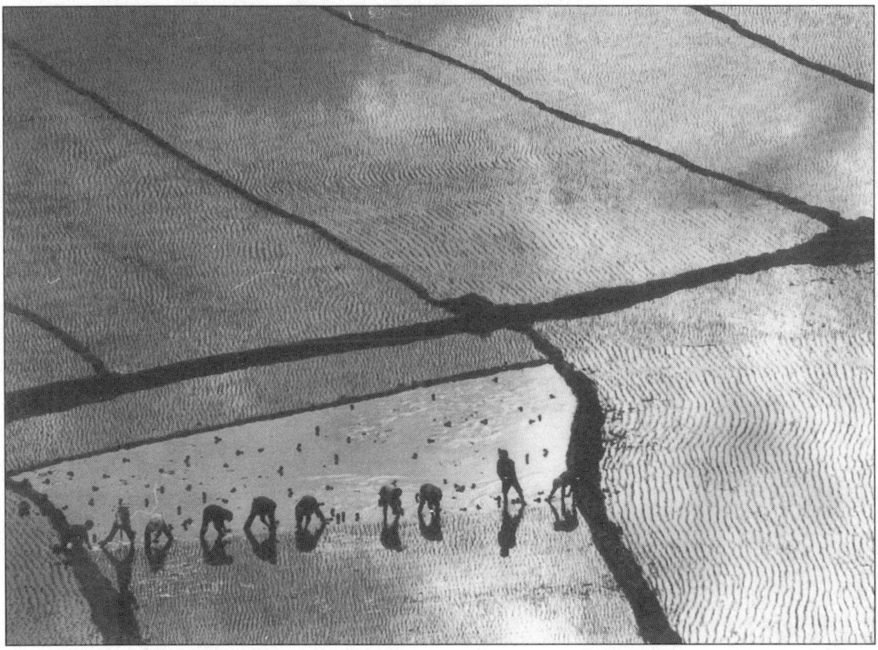

Gao Guangde

An Agricultural Field

Gao Guangde

Working the Land

Wang Xiao

Planting Trees

Gao Guangde

Peasants

Liang Wenjun

Fuel Forest

Gao Guangde

Xi Zang Autonomous Region

Gao Guangde

Grazing Land

Gao Guangde

Peasant Women

Liang Wenjun

Family Stove

Gao Guangde

Coking

Gao Guangde

Agriculture and Industry

Gao Guangde

Urban Scene

Gao Guangde

Environment-Monitoring in Shen Zhen City

Gao Guangde

Gezhouba Dam

Gao Guangde

Wastewater Treatment Plant

Gao Guangde

City Scenery

Gao Guangde

Enjoying the Benefits of Family Planning

Gao Guangde

School Children

Gao Guangde

Children Draw Pictures to Celebrate Environmental Day

Gao Guangde

Qu Geping, former administrator of the China National Environment Protection Agency, accepting an award at the Brazil Environment and Development Meeting, 5 June 1992

9

Population and Water Resources

Water is a precious commodity that is also in high demand. Improved living standards, a growing population, and economic development each contribute to this demand. If China is to continue along its stable path of social and economic development, it will have to bring water shortages under control.

Essential to the survival and development of every organism, water is an integral component of life on earth. It regulates both bio- and geochemical cycles in a dynamic biosphere. Energy transfer and material circulation of all sorts depend on water. Circulating from land to ocean to sky and back, water is a renewable resource. Its aggregate supply is, therefore, less of a problem in relation to increasing rates of consumption, contamination, and redistribution than nonrenewable resources.

Organisms on land are dependent on water, the medium for all metabolic activities. Among other things, water releases heat and maintains constant body temperatures through evaporation. An adult male of 60 kg weight, for example, emits approximately 1,000 ml of water and 539 calories of heat per day by breathing and sweat evaporation. Water further helps organisms assimilate nutrients and discharge excrement.

On a larger scale, water is vital to industrial and agricultural activities. Processing 1 ton of steel necessitates 18–25 m^3 of industrial water consumption; manufacturing a car requires 46 m^3; processing a ton of paper requires 130–165 m^3; viscose yarn filament requires 656 m^3; a ton of rayon requires 2,000–2,100 m^3; a single generator thermal power station of 60 kW requires 2,945–3,926 m^3; and each ton of vitamin E requires 360,000 m^3 to produce commercially. In North China, growing vegetables on 1 ha requires 1,390–1,950 m^3 of water; a wheat field requires 2,224–2,780 m^3.

To protect the environment and maintain ecosystems, water in rivers and lakes needs to be kept at a certain level to meet the needs of fish and other aquatic organisms, to wash away excess salt from fields, and to perform other invaluable functions.

Archaeological and anthropological research suggests that hunters and gatherers were very particular about their surroundings. Habitation frequently occurred in dry, sunny, and elevated areas close to a water supply. The Nile, the Yellow River, the Ganges, and the Tigris are all birthplaces to early civilizations. These rivers each attest to the aforementioned theory.

As people's living standards improve, per capita demand for water tends to rise. It is estimated that humans consumed 12 liters of water per day in the years before the Christian era, 20–40 liters per day in the Middle Ages, and 60 liters per day in the eighteenth century. In urban areas today, humans may use as much as 500 liters of water per day and 104 m^3 per year. Water consumption worldwide is increasing at an annual rate of 80–90 billion m^3. Humankind is clearly becoming increasingly dependent on water. Annual consumption of water, at 3 trillion m^3, is far higher than that of other resources.

According to a projection by the United Nations Food and Agriculture Organization (FAO), the per capita availability of water will decrease by 24% by the year 2000 because of population growth. Water shortage will become more frequent in a number of regions by 2000. At the same time, increases in water consumption will increase rates of contamination. In 1970, approximately 400 billion m^3 of polluted water was discharged; by 2000, this figure is expected to increase by a factor of 14.

Current Affairs

Average precipitation exceeds 6 trillion m^3 per year, amounting to 650 mm in average precipitation depth over the entire country of China. Distribution is uneven, however, with greater amounts in the Southeast and less in the Northwest. Mountainous areas tend to draw more precipitation than do the plains. Annual total runoff in China is 2.7 trillion m^3, or 5.1% of a world total 47 trillion m^3. China ranks fifth in aggregate runoff after Brazil, Canada, the United States, and Indonesia (see Table 9.1). Average per capita runoff is 2,400 m^3 per year, or one-fourth of the world's average, one-fifth of the U.S. average, one-seventh of Indonesia's average, and one-fiftieth of Canada's average. China's per capita water availability is three-quarters that of Japan in spite of the fact that Japan's aggregate water resources are only one-sixth those of China. Runoff from cultivated areas in China is 1,900 m^3, less than the world average.

Groundwater resources amount to 871.68 billion m^3, of which 674.38 billion m^3, or 77%, is in mountainous areas and 187.20 billion m^3, or 22%, is in the plains. Extractable groundwater resources amount to 294.3 billion m^3 per year, or 34% of the total. In plains areas, extractable water resources amount to 187 billion m^3 per year, or 74.7% of existing groundwater supply. In mountainous areas, extractable water resources reach 107.3 billion m^3 per year—only 5.9% of natural groundwater resources in the areas. Total extractable groundwater nationwide in China amounts to 158.9 m^3 per year, or 54% of the country's

Table 9.1 Total Annual Per Capita and Per Mu Water Runoff

Country	Annual Runoff (trillion m^3)	Population (100 million)	Runoff per Person (10,000 m^3)	Farmland (100 million ha)	Runoff per hectare (10,000 m^3)
World total	47.00	43.35	1.08	138.90	0.24
Brazil	5.19	1.23	4.22	4.85	1.07
Canada	3.12	0.24	13.00	6.54	0.48
United States	2.97	2.20	1.35	28.40	0.11
Indonesia	2.81	1.48	1.90	2.13	1.32
China	2.71	11.30	0.24	14.13	0.19
India	1.78	6.78	0.26	24.70	0.07
Japan	0.42	1.16	0.36	0.65	0.65

total supply. Areas of the plains that are suited to groundwater extraction lie in the North where there is scant concern for runoff. The region is accountable for only 17% of the country's total runoff.

A unique geography and monsoon climates help characterize China's water resources. Seasons vary markedly in precipitation, with light rainfall in the winter and spring and frequent rains in the summer and autumn. Rainfall from June to September accounts for 85% of annual rain in the North and 60–70% of annual rain in the Southeast. Frequented by storms, rivers typically rise rapidly and water surges into the sea. When rivers are incapable of bearing runoff, floods occur. Precipitation can vary tremendously from year to year. Annual runoff in the Yellow River Valley, for instance, decreased from 7.19 billion m^3 in 1921 to 2.7 billion m^3 in 1978. Rivers commonly sustain periods of several years of drought or flood. The Songhua River in Jilin Province, for example, recently underwent an 18-year-long dry period and then a 7-year-long wet period.

Water resources throughout the country are unevenly distributed. Precipitation decreases from 1,800 mm in Southeast China to 200 mm in the Northwest. Annual runoff from the Yangtze, the Zhujiang, and other rivers in the South accounts for 82% of the country's total runoff. Farmland in those areas, however, accounts for only 38% of the national total. Annual runoff from the Yellow, the Huaihe, and other rivers in the North adds to less than 18% of the country's total. Farmland in these areas, however, accounts for 62% of the national total. Dry areas with annual precipitation less than 400 mm cover 45% of the territory. Water shortages frequently affect Beijing, Tianjin, Dalian, Qingdao, and other northern cities.

Distribution of groundwater resources is also unbalanced. In mountainous areas, average groundwater deposits vary between North and South China by a factor of 4.2. Deposits average 165,000 m^3/km^2 in the South and 39,000 m^3/km^2 in the North. In some regions of China, the global climate has had a pronounced influence on groundwater depletion. Approximately 30,900 km^2 of glacier,

3,000 billion m^3 of water reserves, and 25 billion m^3 of snowmelt supply water to inland river systems throughout the Northwest. Increasing temperatures in the area, however, have increased glacier melt and reduced reserves by one-fifth every 20 years. After a certain amount of time, snowmelt falls off sharply as does runoff in the river systems. Rising temperatures meanwhile intensify evaporation and water supplies dwindle further.

Water's chemical traits are a function of location and pollution in its path. Mineral contents of rivers generally increase from the Southeast to the Northwest. Concentrations in the North average 300 mg per liter. Humid areas in the Southwest have concentrations as low as 50 mg per liter. Water hardness increases along this same gradient. Annual ion runoff in China's rivers totals 419 million tons. An estimated 43.6 tons of salt per km^2 is lost annually. Silt clouds China's rivers and averages 3.5 billion tons annually. An estimated 115 rivers carry in excess of 10 million tons each of silt per year. The Yellow River carries 16 billion tons per year and its runoff totals 54.3 billion m^3. Rivers clouded with silt will pick up other contaminants as well. Banks also become more gentle as the river fills in.

Overuse

Since 1949, improvements in sanitation have decreased rates of infant mortality, increased birthrates, and prolonged life expectancy. An expanding population has helped cause serious water shortages and widespread pollution. Population growth and poor distribution have, in turn, become root causes to a whole host of problems.

China's population has more than doubled in 40 years, from 0.5 billion in 1949 to 1.1 billion in 1990. This increase suggests that per capita water availability has been cut in half. Aside from the jump in number of people, however, affluence has also increased use of water. Improved living standards, urban population growth, and economic development combine to increase per capita household and industrial water consumption. This has caused widespread shortages. In Beijing, for instance, gross agricultural and industrial output reached RMB 35.91 billion yuan in 1987, by which time the city's population had ballooned to 10 million. Population densities of the 4 principal urban districts reached 26,000 people per km^2 and the aggregate urban population rose from more than 1 million to more than 5 million.

Successive years of drought have damaged water supplies nationwide. In Beijing alone, reserves have fallen sharply in 80 reservoirs, including those at Guantin, Miyun, Shi Sanlin, and Huairou. Reserves were extremely low in 90% of Beijing's reservoirs during the summer drought of 1983. Water shortages reached as high as 100,000 tons per day during peak consumption periods. Some 353 enterprises, including hospitals, suffered from the shortages. Hospital orderlies spent time fetching water in buckets during surgical operations.

In Shandong Province, the population far exceeds water availability. Per capita water resources are only 17% that of the national average. The province lacks an estimated 9–12 billion m^3 of water every year. Water for daily needs and urban industry amounts to 1.3 billion m^3. Irrigated areas total 3.714 million ha and account for only 83% of arable areas in need of irrigation.

Average per capita water consumption in urban areas of China amounts to 90 liters per day. Consumption in large cities averages 100–150 liters per day, whereas consumption in small and medium-sized cities averages 30–70 liters per day. Northern areas generally use less water per capita per day because of their relatively constricted supply. By comparison, people in large cities of other countries consume an average 200–300 liters of water per day.

Irrigation works throughout the country currently have a capacity to store 426 billion m^3 of water. An estimated 29% of this figure comes from large and medium-sized reservoirs, 25% comes from smaller reservoirs, and 46% from rivers. Because river water is not as easy to regulate, this source is less reliable. Approximately half of the currently cultivated areas still depend on rain and snow for irrigation. An estimated 93.3 million ha of grassland is currently water deficient, and 20.0 million ha of fields experience drought annually.

Combating drought has been an ongoing struggle for many of China's cities, provinces, and regions. In the farming and pastoral areas of the Northwest, approximately 40 million people and 30 million livestock are impacted by water shortages. Demand in urban areas grows at similarly rapid rates because of improved living standards and the recent influx of rural peasants. Water projects simply cannot keep pace. In 1979, an estimated 154 cities across the nation survived water shortages totaling 8.8 million tons per day. By 1984, these figures grew to 192 cities and 12 million tons per day. Various industrial and agricultural projects in the Yellow River Valley applied for a volume of water several times that which flows down the Yellow River. Many of these same projects currently operate under capacity or have ceased operation altogether.

Contamination

Water shortage problems are further compounded by contamination, inefficient use, and high water quotas. The volume of water used in agriculture is unusually large. Backward irrigation methods include flooding and are generally inefficient. More than 15,000 m^3 of water is required to flood a single hectare of cultivated land annually; this is twice the amount of actual water requirements. Flood irrigation commonly consumes 30% more water than spray irrigation and 70% more than trickle irrigation. Irrigation systems are also commonly incomplete, poorly built, and hampered by seepage. Large and medium-scale irrigation projects effectively irrigate less than 70% of designated irrigation areas. Utilization of water in channels is below 50%. Preliminary analysis shows that seepage loss amounts to more than 170 billion m^3 per year, or 56% of the channels' total water. As in urban areas, water projects for agriculture simply cannot keep pace with changes brought about by rural economic reforms.

Poor recycling of industrial wastewater also contributes to water shortages. Foreign enterprises have recycling rates as high as 70–80%, whereas China's enterprises stagnate at 50–60%. In Shanghai, for example, where recycling rates are high relative to other areas, an estimated 47% of polluted water was recycled in 1984. This is lower than in Japan (48.1%) and the United States (49.0%).

Underground water systems in urban areas are also in dire need of repair. Seepage rates are generally 5–10% and as high as 15% in some areas. Most water pipes were laid before the 1950s and have undergone patchwork repair. If seepage rates could be reduced by 1%, the amount of conserved water would equal the annual requirements of the population in a large city, or a water project capable of supplying several hundred thousand tons of water per day.

Poor Pricing

A poor pricing mechanism is a primary cause for inefficient water use. The cost of water currently incorporates only 1% of actual production costs. Too many enterprises are more willing to pay higher prices than they are to build water recycling facilities that conserve water and save money in the long run. In spite of the small amount, a uniform water fee is in fact collected throughout Liaoning Province. Groundwater fees are RMB 0.12 yuan per ton for industry, RMB 0.045 yuan per ton for people's livelihood, and RMB 0.016 yuan per ton for agriculture. Surface water fees are RMB 0.03 yuan per ton for industry, RMB 0.01 yuan per ton for people's livelihood, and RMB 0.001 yuan per ton for agriculture. The price of water is generally below cost because it fails to account for discharge, disposal, and forgone future opportunity costs.

Dropping Water Tables

In 10 northern provinces, municipalities, and autonomous regions, overextraction has caused water tables to drop in more than 50 locations. Groundwater levels have dropped 10 m on over 30,000 km^2. Funnels, or cones of depression, will subsequently form, at the center of which water tables drop 10–30 m. The cumulative drop in water at the center represents 10–40% of the average depth of water-bearing strata. Average water levels drop at rates of 1–2 m per year, and some as much as 4.3 m per year. In some areas, the water-bearing strata are entirely dry.

The suburbs of Beijing encompass 100 km^2 and include 6 water projects and 2,000 self-owned wells for industry. An estimated 850 million m^3 of groundwater is extracted annually. Average annual replenishment, however, is only about 600 million m^3. A 1,000-km^2 cone of depression has hence formed under Beijing and attests to overextraction throughout the years. The water table currently drops more than a meter per year. In some areas it has dropped more than 40 m in four decades. All available surface water has been used without additional inputs. Extraction of groundwater reserves is easing, however, by 5% per year.

In Hebei Province, surface water runoff fell from 6.9 billion m^3 to 3 billion m^3 recently. Groundwater levels, furthermore, drop at a rate of 1 to 2 m per year. An estimated 30 cones of depression have accompanied the drops and cover more than 14,000 km^2. An estimated 500,000 motor-driven wells compete for water from the aquifer. Approximately 5%, however, are dry and another 40% produce but a trickle. The water table in this area has dropped from 33.0 m in 1976 to 75.7 m in 1986. Where cones of depression overlap, interconnect, and expand, sink holes may develop, causing damage to buildings and personal property. In Tianjin alone, an estimated 7,300 km^2 of ground, or 64% of the city's total area, has subsided to some degree. Sinking averages 85 mm per year and has been as high as 32 m in some areas. In 1980, the ground in Tianjin subsided by 291 mm.

A clear boundary exists between fresh and salt-laden groundwater near the coastal areas. Seawater, however, will generally fill in drained freshwater aquifers. Encroachment by seawater tends to contaminate inland water supplies and make land less productive. In Shandong Province, the coastline runs more than 3,000 km. An estimated 863 km^2 of land near this coastline has been infiltrated by seawater since 1975. Mineral concentrations, and especially salt contents, have increased dramatically. In some areas, mineral concentrations have jumped from 0.45 mg/liter to 5.5 mg/liter; salt concentrations will reach 46 mg/liter. Water for irrigation and drinking hence becomes increasingly scarce.

Constricting Surface Water Supply

Population growth can strongly impact a country's surface water resources. Among other things, a burgeoning population will drain lakes and streams to reclaim land for cultivation. Throughout China's history, lakes have contributed to the country's economic construction and people's livelihood. Population growth and errors in government policy, however, placed undue emphasis on grain production. Naturally vegetated areas were destroyed and soils eroded. Silt subsequently collected in lakes and reservoirs. It became commonplace to engage in such activities to the detriment of China's natural water resources.

Land reclamation after 1949 claimed more than 1.33 million ha of lakes across China, and 35 billion m^3 of freshwater. The aggregate number of lakes fell from 2,800 to 2,350, a 16% decrease. Poyang Lake, China's largest freshwater lake, has shrunk by 40% since the early 1950s. The Gan River facilitated degradation by carrying 21 million tons of silt into the lake. Its banks currently rise 2–3 mm per year. By 1978, the lake had collected 6.45 million m^3 of silt.

According to the annual *Report on the State of the Environment*, discharge of polluted water (excluding towns and township enterprises) totaled 35.4 billion tons. An estimated 24.9 billion tons is industrial waste discharged without treatment. Rivers in 65 of 94 cities, or 69.1%, are contaminated to varying degrees. Primary pollutants include ammonia, nitrogen, volatile phenol, and oxygen-consuming organic substances. Groundwater quality in most cities has deteriorated because of point pollutants. The primary pollutants include nitrate, sulfate, chloride, and phenol. The following cities are a subsample of 54 cities tested that pass groundwater quality standards: Xining, Lanzhou, Ji'nan, Qingdao,

Nanjing, Hangzhou, Zhengzhou, Changsha, Nanning, Liuzhou, Guilin, Guangzhou, and Kunming. Harbin and Xi'an are cities with badly contaminated groundwater.

Pollution crises are frequent, widespread, and dangerous. An estimated 2,573 pollution-causing accidents occurred in 1983, and 3,617 accidents occurred in 1987. The annual rate of increase in number of accidents per year is 10%. Pollution in 1987 triggered breakouts of hepatitis in Shanghai, infecting 310,000 people. Disease incidence grew to 4.8%, at which point school classes were suspended and factories closed. Needless to say, Shanghai's economy suffered disastrous consequences.

Future Shortages

By the year 2000, water requirements will increase to 20 billion m^3 for urban household use and 130 billion m^3 for industry. To ensure that grain output reaches new highs, field irrigation will also expand from 49 million ha to 53 million ha. Water requirements for the countryside will total 560 billion m^3, so that the total national demand will sum to an estimated 710 billion m^3.

At present, the national demand is 470 billion m^3, or 16% of China's aggregate extractable water supply. If the demand grows as estimated, by the year 2000 China will be using 25% of its total supply. Utilization rates do not appear to be a problem. Hidden beneath these figures is the far more serious dilemma of unequal distribution. In the North, use of surface water has reached 43–68% and use of groundwater ranges from 40% to 80%. China's current financial state and its limited investment capability make it improbable that the water supply will expand. By 2000, the storage capacity of China's water projects will amount to 670 billion m^3. This leaves a deficit of 42 billion m^3, which will have adverse effects on industrial and agricultural development. Unless further measures are taken, pollutant discharge will increase in pace with population growth and economic expansion. Water quality near cities will further degrade, and shortages will become more acute.

Coordinating Population Policies and Water Resource Management

China's current baby boom will add at least 15 million people to the population. In 1989, the population exceeded 1.1 billion. It is expected to reach 1.3 billion by 2000 and 1.6 billion by 2050. Economic expansion and population growth individually impact water resource consumption. When the percentage of industrial output in GNP outstrips other sectors, demand for water will rise at even more rapid rates. To best coordinate the relationship between population control and water resource management, China needs to continue current family-planning initiatives, increase public awareness, and encourage conservation and further exploration. China should specifically address the following issues: a "water-saving" plan, revised accounting methods, improved water management techniques, advanced technology, and reformed macropolicies.

"Water-Saving" Plan

Water is essential to socioeconomic development. Along these lines, both exploitation and conservation need to be reconciled. Humans must learn to efficiently use water in industry, agriculture, and daily life. Establishment of a water-saving plan will help achieve this goal.

Water crises are bound to occur north of the Yangtze River. Historical overuse and the region's climate make this area one of the most challenging areas in the country for water management. Unless effective water conservation measures are soon implemented, crises will immediately impact social stability and development trends. South of the Yangtze, water is not in short supply. Droughts do occur, however, in certain areas in the winter and spring. China's approach to overcoming shortages needs to emphasize rational use and pollution prevention.

Water Resource Accounting

Accounting for water resources requires both a quantitative and qualitative inventory of current resources. A pricing system also needs to be established that takes into account varying degrees of shortage and the laws of supply and demand. The rights and duties of ownership and use should also be more clearly defined. Sound accounting may also help establish just compensation for overuse. China should establish water resource accounts for different regions and then integrate them into the national economic accounting system. Depletion of water resources will thus be reflected in national accounting, and uncontrolled use will become less frequent.

Methods in Water Resource Management

China's water resource management is presently in disarray. Although authorities advocate conservation, public enterprises consume and pollute in excessive amounts. The government, therefore, should enforce water resource management, define property rights and duties, exercise sectional control over industry, establish interregional coordination, draw up comprehensive guidelines for use, and promote additional legislative initiatives.

China needs to focus on three management tasks in particular: (1) define overuse, (2) pinpoint inefficiencies and sources of contamination, and (3) involve the public. First, the government needs to strictly regulate overuse by, for example, defining water tables below which all extraction draws to a close. Major users should be registered and technically instructed on degrading practices.

Administrative departments should also use the media to inform citizens about water-level changes, water usage, and conservation efforts. These efforts will enhance public awareness and promote rational use of groundwater resources. In coastal cities, management should recharge empty aquifers. These measures should at the very least stave off seawater intrusion.

Second, use of water resources should be decentralized and made in accordance with local law. Decentralization is essential to enforcement. The relevant departments should assess unique characteristics in China's various regions and formulate policies to best match the regions' conditions. From 1987 to 1990, water management in Beijing reduced leakage, improved general efficiency, and conserved an estimated 17 million m^3. The success is attributed to eight regulations on water use.

The practicality of China's objectives and its water management legislation will determine compliance and enforcement. Successful policies in Beijing reveal a strong institutional framework and should help guide other regions of China. Two three-tiered management networks help to implement policy. One includes the neighborhood, city, and county levels. The other includes factory, department, and city authorities. More than 10,000 full- and part-time water conservation public employees play a key role in public relations, organization, and supervision. The proportion of units that waste water has fallen from 50% to 20%.

Lastly, the government should start a campaign that encourages ordinary people to participate in water conservation. An ample and clean water supply can be guaranteed only if people understand, support, and participate in government initiatives. The public needs to understand that water is, in essence, a limited, nonrenewable natural resource because of unequal distribution. They need to share in the sense of crisis, urgency, and responsibility. Water management authorities at all levels can help by educating the public about the water quality of China's major river systems. Moreover, they can help to introduce and popularize advanced technology in water conservation.

Advanced Water Technologies

At present, China's industry consumes and discharges massive amounts of contaminated water. Reducing consumption and pollution depends on proper administration, adequate investment, and technological progress, among other things. Although China currently lags far behind other countries in both per capita water supply and consumption, the aggregate volume is tremendous relative to the country's supply and distribution. Presuming that China will continue to expand economically, current water-use methods and technologies are inappropriate. The rate of effective investments into China's 5,556 water treatment facilities is a low 31.3% because of depreciated, inoperative machinery.

The government needs to facilitate scientific research and technological advances for water conservation and pollution prevention. Moreover, the government needs to evaluate, prioritize, popularize, and implement current achievements. Deadlines should be established to eliminate backward technologies that consume and pollute excessive amounts of water.

Macropolicies in Water Resource Management

Government spending is limited and massive increases in investments into water resource management are improbable. Changes, therefore, depend on reforms to financial administration of resources, changes in operations and management, and general strengthening of government policies.

As with other natural resources, a more accurate pricing system is essential. More accurate prices will both restrain contamination and reduce inefficiencies. Price reforms should account for opportunity costs, supply and demand, and development circumstances. A rational pricing mechanism will allow water supply projects to expand while rates of pollution decrease and rates of conservation increase. China is still dependent on agriculture and unable to make sweeping changes overnight. Uniform water pricing for households, industry, and agriculture across China would simply be unbearable. The government will need to subsidize prices according to use, especially with regard to agriculture. Farmers, for example, will pay a price subject to the local economy and government subsidy.

Funding for large water projects must come from bank loans, not from government. Individual, collective, and foreign sources might be solicited to raise funding as well. The state can serve to guide joint ventures in each of these sectors. In this way, the government will refrain from "free" handouts. Individual projects will be forced to improve efficiency and accountability or face foreclosure. This should help to prioritize, streamline, and quicken the development of various projects.

Socioeconomic benefits should also increase. In a commodity economy, water supply enterprises are bound to seek the most profitable markets. Government policies should be established subject to this natural law. On the condition that water supplies to agriculture are guaranteed, the government should lift policy restrictions and grant water supply enterprises the right to select their consumers. This arrangement epitomizes coordination of government interventionist policies and free market forces in the supply and distribution of water resources. New water sources will emerge where profitable water conservation measures become attractive.

10

Population and Energy Resources

Population growth and the availability of energy resources are closely interrelated. Inefficient use of China's energy resources has caused shortages and environmental pollution, thereby infringing on people's general livelihood. Coordination of population policies and energy resource management is important for all nations and China in particular.

Classification

Energy can be categorized in a number of different ways. Two forms include kinetic energy (e.g., moving water or wind) and potential energy (e.g., coal, petroleum, natural gas, nuclear). Energy can further be classified as primary or secondary. The former is energy collected in its natural state. The latter is energy obtained after processing of some sort. The types can be difficult to distinguish.

Energy is further classified as conventional or new, according to its role in contemporary society. For example, conventional energy commonly refers to energy used frequently in daily life. New energy, on the other hand, might refer to energy sources that are under research and experimentation for universal application. Energy is perhaps most importantly also classified as renewable and nonrenewable. Renewable energy typically refers to primary energy sources that are continuously regenerated by nature, presuming sustainable extraction. Nonrenewable energy typically refers to energy whose formation dates back to prehistoric ages. Sustainable extraction, therefore, is possible but on an unrealistic time scale. Fossil fuel deposits typify a nonrenewable energy source because their formation resulted from excess energy reserves thousands of years ago. At current rates of extraction, these deposits will deplete in a relatively short time frame. Fossil fuels are currently the world's primary source of energy.

Energy and Economic Development

As with population growth, energy consumption is also closely related to economic development. The ratio of annual growth of energy consumption to annual growth in GNP is commonly referred to as the elasticity coefficient of energy consumption. This coefficient is affected by factors such as economic organization, utilization rates of energy resources, technological advances, administrative efficiency, and consumption. At initial stages of industrialization, elasticity coefficients are typically greater than 1. In other words, the growth rate of energy consumption is greater than the growth rate of the national economy. Economic reconstruction, administrative streamlining, technological advances, and increased efficiency should decrease the coefficient so that the growth of the economy outstrips growth of energy consumption. In developed countries, elasticity coefficients of energy consumption generally equal 1. Between 1949 and 1980, China's elasticity coefficient averaged 1.17. This figure incorporates commercial energy consumption, and industrial and agricultural productivity for the numerator and denominator, respectively. Economic reconstruction and energy conservation campaigns in recent years have helped to deflate the coefficient.

Improved living standards typically increase the demand for material products and energy consumption. Conversely, annual per capita energy consumption is commonly regarded as a key indicator of people's living standards. As demonstrated in Figure 10.1, increases in per capita energy consumption directly correlate with per capita growth in GNP. The world's aggregate energy consumption has rapidly increased. Increases in coal, crude oil, and natural gas production during the mid-1970s exceeded population growth. Consumption varies widely, however, from country to country. In 1975, per capita energy consumption in the United States was 11 tons of coal equivalent (t.c.e.), and per capita consumption of electricity was 9,370 kWh. Both figures are the highest in the world. China's per capita energy and electricity consumption was considerably lower at 0.49 t.c.e. and 213 kWh, respectively. In 1975, developed countries accounted for 25% of the world's population and consumed 80% of the world's energy. In 1980, the United States accounted for 5% of the world's population and consumed 25% of the world's energy. It consumed 49% of the world's natural gas production, 25% of the world's crude oil production, and 20% of the world's coal production. Conversely, India accounted for 15% of the world's population and consumed 1.5% of the world's energy. In 1980, per capita energy consumption in the United States was greater than that in developing countries by a factor of 54, in the world by a factor of 8, and in industrialized countries by a factor of 2 or 3.

Per capita energy consumption in China has been comparatively low at 0.60 t.c.e. per capita in 1977, 0.63 t.c.e. per capita in 1981, and 1.04 t.c.e. per capita in 1990. Per capita consumption in Britain, West Germany, and Japan are far higher at 12.3, 6.3, and 4.7 t.c.e., respectively. The relationship between

Figure 10.1 Per Capita Energy Consumption Versus Per Capita GNP

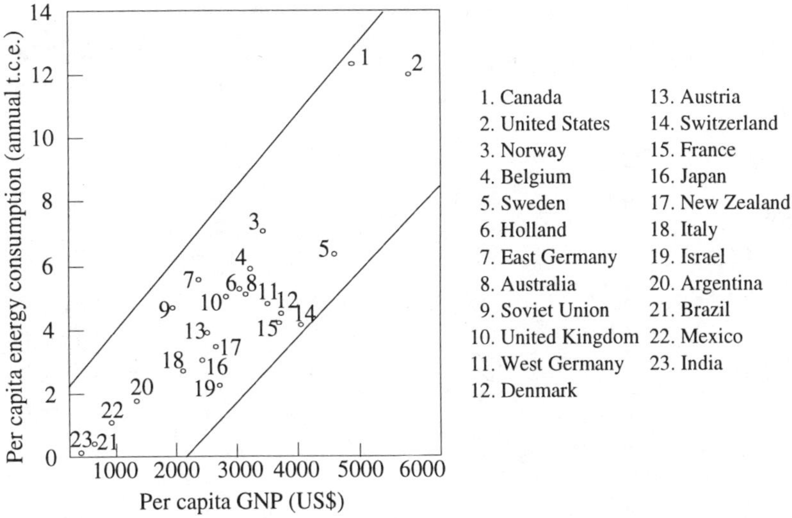

population growth and energy consumption is not simply linear, then, when viewed in a global context. In developing countries with large populations and rapid population growth rates, however, the influence of population growth on energy consumption is clear and obvious.

Current Affairs

Fossil energies are the most widely used energy resources in the world. In 1979, petroleum comprised 44% of the world's energy consumption, replacing coal as the leading form of energy. Petroleum typically replaces coal as a country industrializes. In Japan, for example, petroleum accounts for 75% of total energy consumption. Use of natural gas is also becoming increasingly attractive because of both convenience and limited pollution.

China depends primarily on coal as its primary source of energy (see Tables 10.1 and 10.2). In 1990, coal consumption accounted for 75.6% of total energy use. The remaining resources include petroleum (17.0%), hydroelectricity (5.8%), natural gas (2.1%), and nuclear energy (0%). In recent years, people have become increasingly aware of the limited availability of fossil fuel deposits. Population growth will quicken the depletion of these nonrenewable energy sources and increase atmospheric pollution. Future energy supplies and the status of our environment are in jeopardy until cleaner alternative energy sources are made affordable.

Table 10.1 Energy Production in China, 1978–1990

Year	Total energy production (10,000 t.c.e.)	Percentage of total energy production: Coal	Crude	Natural gas	Hydroelectricity
1978	62,770	70.3	23.7	2.9	3.1
1979	64,562	70.2	23.5	3.0	3.3
1980	63,735	69.4	23.8	3.0	3.8
1981	63,227	70.2	22.9	2.7	4.2
1982	66,778	71.2	21.9	2.4	4.5
1983	71,279	71.6	21.3	2.3	4.8
1984	77,855	72.4	21.0	2.1	4.5
1985	85,546	72.8	20.9	2.0	4.3
1986	88,124	72.4	21.2	2.1	4.3
1987	91,265	72.6	21.0	2.0	4.4
1988	95,018	72.9	20.6	2.0	4.5
1989	101,639	74.1	19.3	2.0	4.6
1990	104,000	74.2	19.0	2.0	4.8

Source: *China Statistics Almanac,* 1991.

Table 10.2 Energy Consumption in China, 1978–1990

Year	Total energy consumption (10,000 t.c.e.)	Percentage of total energy consumption: Coal	Crude	Natural gas	Hydroelectricity
1978	57,144	70.7	22.7	3.2	3.4
1979	58,588	71.3	21.8	3.3	3.6
1980	60,275	72.2	20.7	3.1	4.0
1981	59,447	72.7	20.0	2.8	4.5
1982	62,067	73.7	18.9	2.5	4.9
1983	66,040	74.2	18.1	2.4	5.3
1984	70,904	75.3	17.4	2.4	4.9
1985	76,682	75.8	17.1	2.2	4.9
1986	80,850	75.8	17.2	2.3	4.7
1987	86,632	76.2	17.0	2.1	4.7
1988	91,500	76.0	17.2	2.1	4.7
1989	96,934	76.0	17.1	2.0	4.9
1990	98,000	75.6	17.0	2.1	5.3

Source: *China Statistics Almanac*, 1991.

China's coal deposits and hydroelectricity potential rank among the largest in the world. Total coal deposits buried less than 1,500 m below ground amount to an estimated 4 trillion tons. By the end of 1989, coal deposits confirmed by exploration totaled 901.5 billion tons and raw coal output amounted to 1.054 billion tons. Potential hydroelectricity is estimated at 676 million kW, of which 378 million kW may currently be used. Average annual output of electricity is 1,920 billion kW, ranking first in the world. China also has abundant nuclear energy resource potential. Preliminary stages of the Zhejiang Qinshan Nuclear Power Plant have begun, and the Guangdong Daya Bay Nuclear Power Plant is under construction as planned.

China is also abundant in petroleum deposits. Oil and gas deposits cover 3.20 million km^2 of the land and contain 30–100 billion tons of oil. At sea, oil and gas deposits cover an estimated 1.4 million km^2 and contain 5.3 billion tons of oil. Annual output of crude oil is 137.6 million tons, ranking fifth in the world.

China's fuel forests are another prime source of fuel for cooking and heating in the rural countryside. Approximately 2 million ha of planted fuel forests in addition to shrub woods, production forests, and shelter forests currently provide rural areas with 88.6 million tons per year of wood for fuel. China is additionally rich in renewable energy resources such as solar, wind, tidal, and geothermal energies. In spite of their potential, these energy resources require massive investments to harness and exploit. Approximately 160 GW, or 10% of the total potential wind power, is exploitable. Tidal energy resources, located in the Zhejiang and Fujian coastal areas, total 20 GW.

In spite of China's vast energy resources, the country's population makes per capita figures pale in comparison to world averages. Moreover, the distribution of energy resources is far from equitable, and the country lacks adequate infrastructure to overcome geographic distances. For instance, an estimated 70% of China's hydroelectricity is located in the Southwest. Insufficient geological surveys currently limit large-scale water projects in the region. Also, an estimated 60% of China's coal reserves are located in the North. Per capita energy consumption in 1990 stands at 1.04 t.c.e., which is 7–17% that of more developed, industrialized countries. A relatively high elasticity coefficient of energy consumption is bound to fall with energy conservation efforts as planned.

China's use of energy resource deposits dates back 3,000 years with the discovery of coal's attributes. Firewood and straw remained China's primary sources of energy, however, through the end of the last century. Population growth and migration to dry and remote mountainous areas have caused increasing deforestation at unsustainable rates. At the turn of the century, imperialist powers entered China's markets by force and plundered the country's mineral and energy resources. Modern industry since the 1950s has primarily relied on coal consumption. The dependency has not been of benefit to China's environment.

Since 1949, China's energy production and consumption have expanded with population growth and improved living conditions. Since the early 1980s, components of energy consumption have basically remained stable. Coal has been China's principal energy source; noncommercial biological energy (e.g., combustion of straw, firewood, and animal dung) has played a strong secondary role in people's daily lives. New, renewable energy resources continue to be explored.

Energy resources have continued to be in short supply because of China's expanding energy-consuming industries and sustained, low energy prices. The gap between energy supply and demand in China will widen into the next century as the impacts of inefficient energy use on the environment grow more severe. Although only one of many factors, the population needs to be kept under control to deflate China's growing demand for energy resources.

Population growth, improved living standards, and an exponentially increasing GNP all suggest that the demand for energy will intensify in coming years. Two variables concerning energy resource consumption can be presumed constant as China enters the twenty-first century. First, the types of energy consumed will remain the same. China will depend on coal as a primary resource regardless of its quality. Use of coal will pollute the air and land with acid deposition and increase emissions of greenhouse gases. Second, China's population will remain predominantly rural because of strained conditions in urban areas and the rise of township enterprises. Pressures for fuel from forests and grasslands therefore will not lessen in the foreseeable future. If deforestation and grasslands destruction continue at current rates, diverse ecosystems throughout the rural countryside will soon be threatened by extinction.

In spite of differences in how population growth impacts developed and developing countries' per capita energy consumption, population growth has the most pronounced effect on aggregate energy consumption in developing countries such as China. If energy consumption per capita per year is presumed to be 1 t.c.e., aggregate consumption will then increase by 15 million t.c.e. per year. Furthermore, if energy consumption per capita per year is estimated at 1.5–1.6 t.c.e. for a comfortable life, then annual energy consumption will increase by 22,250–24000 t.c.e. When China's population reaches 1.3 billion people by early next century, then an additional 300–320 million t.c.e. will be required to meet the standards of a comfortable life. This is over and above the current 1.95 to 2.08 billion t.c.e. consumed, and far beyond China's energy production capabilities.

Consumption and Pollutant Discharge

Because coal is China's predominant source of energy (see Table 10.1), objectives to expand energy production and control emissions of harmful gases are bound to clash. China's total emissions increased steadily between 1983 and 1990 (Table 10.3). Sulfur dioxide (SO_2) and concentrations of dust increased between 1983 and 1988, dropping somewhat thereafter because of drops in the rate of economic expansion. Gases from fossil fuel combustion, the main source of air pollutants in China, increased at an annual average rate of 6%. More than 90% of SO_2 emissions result from coal burning; the amount of gas emitted increases at an annual average rate of 4.8%. Carbon dioxide emissions are primarily attributed to coal burning as well. In 1989, China discharged 420 million tons of carbon, ranking third in the world. Per capita discharge of the gas, however, amounted to a relatively low 0.4 tons. Substantial increases in such greenhouse gases have contributed to the global warming problem.

Rapid population growth has unquestionably slowed the transition from inefficient to efficient energy sources. Sustained population growth and prolonged energy shortages have increased demands for both commercial and noncommercial energies. Consumers are presented with few options in and out of the marketplace; they commonly resign themselves to inefficient, polluting energies such as coal, grasses, and wood. Coal gas is largely inaccessible, and centralized heating is available only in certain urban districts. Households presented with few other options, therefore, burn bulk coal in small stove ovens. The stoves have poor heat efficiency (10–20%) and cause serious air pollution at low altitudes. The air contaminants present a serious health hazard to urban inhabitants.

When shortages of commercial energy occur, rural inhabitants rely on burning straw, firewood, and animal dung. Subsequent damage to the environment is considerable. Zhejiang Province lost 3.69 million ha, or 43% of its mature forests, in the early 1970s. Other provinces were deforested at rates of comparable severity. The excessive tree harvests destroyed natural vegetation, worsened soil erosion, lifted riverbeds, silted reservoirs, and generally increased the incidence of natural disaster. Protection of root structures is commonly overlooked. If the roots of a tree are dug up by peasants in dire need of fuel, vegetative sprouting by hardwoods is inhibited and soil erosion becomes even more disastrous.

Research by the Lanzhou Desert Research Institute of the Chinese Academy of Sciences has demonstrated that the main cause of desertification in North

Table 10.3 Discharge of Dust, Sulfur Dioxide, and Other Contaminant Gases, 1982–1990

Year	Dust (10,000 tons)	Sulfur dioxide (SO_2) (10,000 tons)	Other contaminant gases (100 million m^3)
1982	1,458	1,275	54,412
1983	1,353	1,200	60,941
1984	1,311	1,243	66,969
1985	1,323	1,303	70,654
1986	1,384	1,250	66,679
1987	1,445	1,412	77,270
1988	1,436	1,523	82,382
1989	1,398	1,564	82,685
1990	1,324	1,495	85,000

China is excessive exploitation of the area's natural resources. Forms of exploitation include reclamation of cultivated areas, animal husbandry, deforestation for fuel wood without regard to regeneration, and water extraction from aquifers.

When large quantities of straw are burned for fuel, nutrients combust and are prevented from returning to the field as fertilizer. Cultivated areas are then in need of fertilizer, soils harden, and farming conditions worsen. In many rural areas, organic substances in the soil have dropped to below 1%. Farmland is thereafter more susceptible to degradation from drought, flooding, soil and plant diseases, and insects. Production therefore falls off and consumers suffer the consequence.

Trends

Population control and energy resource availability will continue to steer China's socioeconomic development well into the next century. A sound understanding of the dynamic relationship between the two issues will help to form and implement effective policies. Trends both at home and abroad will be addressed in this context.

The world's population will reach an estimated 6 billion by the year 2000, and 10 billion by the middle of the next century. The growth will primarily occur in developing countries. Developments in the use of energy resources are far more complicated and site-specific than population growth. As the world's energy resources shift from fossil to nuclear and other renewable alternatives, dependence on petroleum is bound to lessen. Although transportation, agriculture, and chemical industries may carry exemptions, other fields will gradually phase out their demand for petroleum and focus more on alternative sources such as natural gas.

Large reserves of natural gas resources will doubtless play an important role in China's economic expansion by the end of the century. China's coal will also make an important contribution to the world's future energy supply by acting mainly to help stabilize other supplies. Coal transformation (from solid to less damaging gas and liquid states) is therefore essential. Additionally, the development of nuclear power technology has been slower than expected, in spite of its enormous potential, because of environmental and safety considerations.

Hydroelectricity is poised to play a larger role as a principal energy resource once geological maps define limitations for construction. By the first half of the next century, most exploitable water sources will have been impacted by water projects in one way or another. To mention a single pioneering endeavor, controlled fusion technology offers a unique opportunity to achieve a smooth transition from traditional to more modern, renewable energies.

Most developing countries face severe obstacles in shaping policies that aim to both control population growth and conserve energy consumption. The role that population growth plays in developed countries is not entirely clear. After industrial growth, aggregate consumption of energy resources begins to taper so that fluctuations in population levels are negligible. In developing countries, however, continued population growth is compounded by increasing per capita energy consumption and expanding energy-intensive industries. Population growth, above all other factors, has a prominent role in aggregate energy consumption.

An additional commonly overlooked factor inhibiting conservation in developing countries is the role of tradition. An adequate commercial energy supply will not necessarily replace the use of traditional, time-honored noncommercial energy. Conservation efforts might also focus on enhancing utilization rates and improving pollution control of traditional energy supplies and devices.

China's immediate concerns include energy shortages and accounting for externalities, such as pollution. First, by the year 2000, it is estimated that energy supplies will increase a mere 50% whereas GNP will nearly double. Second, an economy that continues to rely on coal as its primary source of energy needs to account for its externalities. Otherwise, China's economic success will smolder in its own atmospheric pollution.

Policies based on science and democracy should empower energy enterprises and avoid large-scale polluting incidents. In coal production, China needs to strengthen state-owned mines while also developing small- and medium-scale ones. The government should also facilitate coal mining in Shanxi, Inner Mongolia, and Shaanxi to more equally distribute energy production. In light of the fact that China has yet to uncover abundant reserves of oil and gas, the government should further fund exploration in the West while maintaining current oil fields in the East. The government furthermore needs to develop its thermal, hydroelectric, and nuclear power resources. New and renewable energies, although accounting for a small percentage of China's current energy structure, justify research and exploration.

Development of energy resources is an important component of China's socioeconomic development. A sustainable, secure, and nonpolluting energy stock will insure domestic growth and stability. Experiences in other countries suggest that per capita energy consumption should reach 1 t.c.e. to meet demands for a comfortable life-style. By 2000, China's population will amount to 1.3 billion people and living standards should reach "comfortable" levels. If annual GNP growth rates in the 1990s average 6%, then annual average primary energy growth rates should be no less than 3%. In this scenario, the total demand for primary energy will be 1.2 billion t.c.e. in 1995 and 1.4 billion t.c.e. in 2000. To reach these objectives, the country will need to actively conserve and efficiently exploit its energy resources.

Coordinating Energy Policies and Energy Resource Management

Controlling population levels, expanding resource availability, and reducing overall consumption are three of nine different approaches to limiting population growth and relieving energy shortages.

Population Levels

The more impoverished and backward a district, the more children are born. Relatively affluent areas, on the contrary, have far less a problem with family planning. Efforts at population control need to focus on poor rural areas and minority districts. The government should also promote policies for maternity care and child rearing. Uneducated, poor segments of the population generally consume more energy and pollute more areas of the country. When other currently industrialized countries produced GNPs comparable to China's, their relative total energy consumption was far lower. Poor education and limited science and technology inhibit China from greater conservation. By not confronting these obstacles, China will be hard-pressed to afford and implement more efficient industries and energy-saving technology.

Exploring for Additional Resource Stocks

In spite of the currently poor utilization rates, coal mining should continue and expand where possible. The industry's objective is to expand by 40 million tons of coal per year throughout the 1990s. Targeting high-grade coal deposits with low sulfur content in China's central and western regions will help to reduce atmospheric pollution. Moreover, mining at increased distances from urban centers in the East will cut down dust pollutants in cities. At the same time, the country needs to develop its network of power lines so that transportation costs of vast amounts of coal can be substituted by the conduction of electricity.

Establishing thermal power plants may help to ease the country's dependence on coal production. An estimated 30 million kW of thermally generated power proposed for the 1990s will save the transportation costs for 100 million tons of coal. Accelerating the construction of hydroelectric plants will also ease reliance on coal reserves. In the Eighth Five-Year Plan, China should increase the number of large and medium-scale hydroelectric generator plants with capacities greater than 10 million kW. Moreover, energy-storing power plants will ease pressures on peak production periods. Exploration of petroleum and natural gas deposits should aim to increase crude oil production by 2 million tons per year.

In the transition from fossil energies to more modern renewable energies, nuclear power is becoming increasingly attractive. China needs to develop its nuclear power potential. The 300-mW Qinshan Nuclear Power Plant, designed and constructed domestically, has begun to produce energy. The 1,800-mW Daya Bay Nuclear Power Plant is presently on a trial run. Presuming that these installations are safe and efficient for the next several years, China will continue to lay a foundation for sustained nuclear power production into the next century. Research on controlled nuclear fusion has recently gained popularity. The country should pay close attention and actively pursue any breakthroughs.

Rural electrification demands special attention. At present, 32 counties and 190 million people are without electricity. Small and medium-scale hydroelectric power plants should help to expand networks of electricity. Even more remote border areas should focus on harnessing wind and solar power. Rural areas are short of conventional energy sources but rich in solar energy. Research has made progress in the field as well as solar energy a reality. Solar houses, for instance, have reformed traditional methods of heating. Solar energy heats and collects in the walls and warms the ambient air. Solar water heaters also are available now as well as heat-collecting aluminum-copper plates. The products generate and store 200,000 m of energy and are comparable to foreign brands.

Five medium-scale solar installations have been constructed for demonstration. Scientists and technicians have also developed Asia's largest solar desiccator, with a heat-collecting area of 620 m. Second-stage polysilicon solar battery components for practical use are now produced on an assembly line, as well as large-scale silicon casting furnaces. Several types of solar ovens have also been developed. The cast iron oven, in particular, is now popularized as a commercial product and exported. The following is a list of achievements in solar technology since the Sixth Five-Year Plan: (1) solar houses increased capacity by 88%, to 150,000 m; (2) solar water heaters increased capacity by 75%, to 315,000 m; (3) solar desiccators increased capacity by 75%, to 315,000 m; (4) solar batteries increased consumption from 200 kW to 1 mW; and (5) solar ovens increased capacity 50%, to 120,000 m.

Wind power is an additional pollution-free power source with great potential. Areas with strong winds happen to be remote, border areas without adequate supplies of electricity. Wind-power generators of centi- and kilowatt classes are currently being developed for household use. The government plans to install 100,000 wind-power household generators in Inner Mongolia. Wind-powered pumps are currently used in rural areas to irrigate fields, breed aquatic plant and animal life, refine salts, improve soil fertility, and water livestock. One machine in particular, the FDG-7, is the most powerful and efficient in China as well as in the world. Used for two-stage pumping of water into a saltern, the unit could save more than 10,000 kW per year and increase production of crude salt by 70 tons. A single unit can also regenerate water for a shrimp farm, up to one hectare in size. Both energy conservation and the increase in productivity are remarkable.

Use of underground, geothermal heat began in China in the 1970s and has rapidly developed. At the Hawaii International Conference on Underground Heat, China ranked as the second-largest consumer of geothermal heat. China has achieved tremendous progress in the monitoring, applied technology, and environmental protection of geothermal heat sources. Underground greenhouses, year-round fish farming, and other breeding technologies are only some of the benefits of geothermal energies. Four research and demonstration facilities throughout China experiment with geothermal energy's productive potential. In Xong County, Hebei Province, for example, chicken egg hatcheries and leather processing are two fields experimenting with this energy source. The number of wells used for these and other industries have since increased from 4 to 9. Aside from the nonquantifiable benefits of environmental protection, harnessing geothermal energies has yielded an immediate economic benefit of RMB 14.1055 million yuan.

Fuel forests are a principal source of energy in mountainous rural areas. China's burgeoning population has reduced the percentage of fuel wood coming from fuel forests. Population pressures are simply too great for the limited supply of fuel wood in specially designated fuel forests. Rural inhabitants have, hence, legally and illegally encroached on forests designated for timber production or protection. Increasing the area of specially designated fuel forests and popularizing more energy-efficient stoves are two effective measures to reduce the fuel wood deficit. The country needs an additional 4.4 million ha of fuel forests to satisfy demand.

Resolving Inefficiency

Making the country's energy resources more accessible is only one part of the equation to battle energy shortages. Improving utilization rates of energy resources is equally important. Overconsumption and waste are nationwide problems. Consumption per dollar's worth of GNP in China is nearly 3 to 4 times larger than that of developed countries and twice as large as most other developing countries. Energy conservation, then, stands to make gross improvements in efficiency. Conservation is complex, however, and incorporates many variables. It is almost more a societal problem than it is a tangible, systematic problem to overcome. The tangible components include almost exclusively technology for conserving and monitoring energy use. Intangible components include reforming product structures and pricing, accelerating research, streamlining administration, and establishing just legislation.

The Sixth Five-Year Plan introduced a plan to conserve energy in China. A relatively well-organized administrative apparatus has since been established. The State Council also confers annually to initiate conservation measures. The State Planning Commission has additionally established an organization to coordinate energy conservation programs at the national level. Government departments and committees in provinces and municipalities throughout China further establish special organs to administer conservation measures. Most major energy-intense enterprises have special offices for energy conservation.

In the mid-1980s, China also drafted Provisional Regulations for the Administration of Energy Conservation and Regulations for Strengthening Conservation of Electricity. The objectives of these regulations are to promote the use of new conserving technologies and to phase out old equipment. Since the early 1980s, China has achieved many of its goals. Energy consumption per RMB 10,000 yuan of GNP has fallen from 13.9 t.c.e. to 9.8 t.c.e in 1988. Consumption of energy-intense products has fallen by two-thirds, saving more than 20 million t.c.e. per year. The elasticity component for energy consumption has, furthermore, dropped to 0.52. These developments suggest recent rapid economic growth with relatively low growth in energy consumption.

In the next decade, China will implement effective measures to conserve energy and reduce consumption with advances in technology, increased efficiency in administration, and public campaigns to reduce waste. Strict conservation measures will alleviate shortages, promote innovation, increase economic profit, reduce pollutants, and protect the environment. Increases in the percentage of coal converted into electricity are of critical importance. The percentage of coal converted into secondary energies such as electricity, coal gas, and coke is approximately 35%; direct combustion accounts for 65%. Conversion of coal into electricity accounts for only 25%, far below the level of 65% in the United States. Generating electricity with coal will increase combustion rates, improve utilization rates, conserve large quantities of coal, and reduce pollutant discharge. Aside from conservation and protection, benefits include higher profits.

Another priority is promoting appropriate technologies where outdated modes of production contaminate the environment. Legislation may help to overcome reluctance among nongovernmental enterprises. Adequate distribution of coal reserves needs to be guaranteed to all geographic regions. Home construction may also help to reduce the demand for coal in heating by using adequate insulation. These are only a few of many measures for the country to follow.

Reducing Urban Energy Requirements

Consumption of energy resources in urban areas accounts for 75% of the country's aggregate consumption. China needs to revise policies that have given rise to inefficiencies and pollution. For example, urban power plants need to focus on generating heat as well as electricity. Shifting from the production of power to the production of heat will help to phase out small, decentralized, backward heating-supply systems. The government also needs to improve modes of combustion and principal energy sources. Three effective policies include regulating use of coal, promoting use of natural gas, and improving boiler technology and efficiency. First, China should ban the combustion of bulk coal and mandate use of molded coal for domestic use. Use of molded coal will conserve the amounts of coal consumed and reduce SO_2 emissions by 50%, dust by 60%, and hydrocarbons by 50%. Second, China should also promote the use of natural gas. By the end of the century, natural gas will have impacted 50 prominent cities in China. Lastly, China needs to phase out inefficient boiler and combustion devices while expanding the use of energy-saving boilers that raise heat efficiencies by 80–85%. If sulfur-fixing agents are added, SO_2 emissions will fall by 8% as will emissions of nitrogen oxides.

Reforming Rural Energy Policies

By the end of the twentieth century, the total demand for energy in China's rural areas will be approximately 533 million t.c.e. Energy for household use amounts to 35% of this total. To relieve energy shortages in rural areas, China should implement the following measures: (1) popularize energy-saving stoves, (2) promote use of natural gas, (3) increase forested areas specially designated for fuel wood, and (4) construct small hydroelectric plants.

Outdated stoves in China, which have heat efficiencies of only 10%, are still in common use. Each year, 400 million tons of straw, 9 million tons of animal dung, and 300 million tons of firewood are burned for heating and cooking. More modern energy-saving ovens will double heat efficiency so that only a fraction of these sources would have to be burned for the same amount of energy. An estimated 120 million households throughout China's rural areas use the more efficient ovens. The government needs to continue promoting these ovens.

China currently contains more than 7 million natural gas operations of varying capacities. Located in areas abundant in natural resources, natural gas supplements electricity and other fuels. To expand the use of natural gas in rural areas, programs need to be integrated with annual plans established by authorities at provincial, regional, and county levels. Moreover, teams of technicians should organize to improve and popularize the technology in the field. If rural areas contain 20 million households that use natural gas (10% of all rural households in China), the country will save more than 20 million tons of straw. If 50% of China's rural households use natural gas, the country will save more than 100 million tons of straw.

China should redouble its efforts to plant and set aside forests as fuel wood reserves. This measure will help resolve energy shortages in rural areas. Additional benefits include more stable climates as well as water and soil conservation. Barren hills and slopes as well as land unsuited for cultivation should be planted with fuel wood. The selection and planting of species with rapid growth and adaptations suited for the environment is of utmost importance.

Lastly, rural areas should use any and all water resources suited for the construction of small hydroelectric power plants. Harnessing an area's water resources is an effective way to ease energy shortages and protect forests from excessive harvesting. Water resources for small hydroelectric power plants (below 12,000 kW) would supply 76,000 mW of power. The actual generating capacity of installed plants currently exceeds 3,000 mW. An estimated one-third of the nation's counties and 40% of all county factories depend on small hydroelectric power plants for their energy needs. To tap the potential of rural areas' river systems, China is currently running a demonstration plan for the electrification of rural areas. The country will launch pilot projects in 100 counties, which will depend primarily on hydropower. China has already established a program for the construction of small hydroelectric power plants in roughly 500 counties. By early 1990, 81 had been built.

In the past several years, rural areas have experienced tremendous progress in energy production. In the 1980s, newly established fuel forests amounted to 1.62 million ha. Moreover, small hydroelectric power plants exceeded 63,000 in number and 13,000 mW in capacity in 1990, with a generating capacity of 38.5 billion kWh. Construction of solar houses is also in full swing in the North's rural areas. The area of these houses increased from 180,000 m^2 in 1989 to 350,000 m^2 in 1990. An estimated 100,000 households use solar ovens and 1 million households use solar water heaters. Approximately 100,000 wind-powered generators supply herdsmen and fishermen with electricity in Inner Mongolia and the coastal areas. Use of natural gas and other alternative nonbiological energy sources has produced energy amounting to 4.6 million t.c.e. for brick, tea, and tobacco production in rural areas.

Pricing

A primary cause for inefficiency and pollution is a poor pricing mechanism. According to the traditional view, natural resources have little face value. Prices for petroleum and coal, therefore, are derived by adding transportation costs and profit, neglecting any intrinsic value. This sort of pricing might be characterized as follows: "High prices for products, low prices for raw materials, and no price for natural resources." It follows that a ton of coal is cheaper than a ton of sand and a barrel of petroleum is cheaper than a bottle of spirits. It is essential that more accurate value theories be applied to natural resources so that energy prices take into account intrinsic values of energy resources.

Property Rights

Poorly defined property rights add to problems associated with poor pricing. As the material basis for socioeconomic development, energy resources are nearly invaluable. The current constitution stipulates that natural resources be owned by the state. For a long time, however, property rights were so poorly defined that trespassing and overuse occurred frequently. If energy resources were genuinely regarded as national assets, and property issues were more clearly defined, many of the country's energy problems might resolve themselves.

Industrial Responsibility

Industry is guilty of not fulfilling its duties. Ownership, as with citizenship, carries with it certain duties as well as rights. Industry has let go of its responsibilities, including the protection, rehabilitation, regeneration, and accumulation of natural resources. There has been a tendency to squeeze productivity out of nature and to give nothing in return. Needless to say, this is detrimental to the environment and a leading cause of petroleum and gas shortages. China, therefore, should acknowledge industry's impacts on natural resources, increase returns to nature, and intensify protection efforts in the future.

Accounting Methods

Lastly, inefficiencies stem from poor natural resource accounting. Current accounting mechanisms consider only national productivity and growth, without considering depletion or degradation of a country's natural resources. This sort of accounting mechanism suggests illusory growth in GNP and indirectly threatens a country's resources by not identifying them. National accounting systems play an important role in guiding socioeconomic development. China should reform its national accounting system so that the stock and flow of natural resources (including energy sources) is well integrated into the government's balance sheets. More rational pricing, better-defined property rights, acknowledgment of industry's responsibilities, and improvement in the country's national accounting mechanisms are effective measures to ensure the sustained use of China's energy resources.

11

Population and the Living Environment

As humankind shapes the environment, so the environment has shaped humankind. Evolution has indisputably established humans as the most powerful members of the environment. In dominance, however, fragility lies latent. History has shown that if humans fail to keep their population growth under control relative to other living and nonliving things, the environment will continue to degrade. This hypothesis is by no means exaggerated or sensational. Nature frequently warns humans of its ruthless reprisal. Humankind needs to become more aware of its relations with other components of the environment. Otherwise, inevitable destruction and degradation will ensue. Our quality of life depends on both economic growth and environmental protection.

Evolution and Human Adaptation

Prior to the Industrial Revolution, humans "passively" adapted to the environment. In spite of population pressures, environmental impacts by humans were localized and frequently recovered by natural regeneration of the sea, air, or land. The Industrial Revolution, however, introduced "active adaptation" to the environment. Full-fledged assaults on the environment inflicted more damage than the earth itself could heal. A comprehensive survey of the history of evolution may benefit relations between humans and their surroundings.

In the age of hunters and gatherers, limitations on human activities included resource availability and high consumption. Resource availability, in particular, depended on types of resource, the ability to regenerate, nutrient value, geographic distributions, and climatic fluctuations. If limitations were great, people would migrate and population densities would drop off.

Approximately 10,000 years ago, agriculture and animal husbandry began to replace hunting and gathering as the major mode of production. Relations between humankind and the environment entered a new stage in which population pressures intensified agricultural production. Demographic changes became less frequent. Human populations increased slowly at 0.007–0.03%. These rates appear negligible in short time periods; after long periods, however, their impacts are more pronounced. Hunters and gatherers, for instance, were initially in small groups, interspersed, and sparsely populated. As populations increased, however, groups began to disintegrate and continuously migrate. In so doing, new areas were settled.

At about this same time, the world's population reached a carrying capacity of 8 million people. Migration and reclamation increasingly had an impact on the environment. Under these circumstances, innovative technologies became the only means to support ever-larger populations without damaging the environment.

Primitive agriculture, hence, began to enhance the land's productive potential. Increasing demand for food and specialization of labor further contributed to the growth and prosperity of agriculture. As farmers specialized, people became increasingly interdependent. Rotation agriculture began to disintegrate approximately 7,000 years ago primarily due to population pressures. When densities increased to a certain level, existing farms failed to meet the demand. But as tools and techniques improved, each laborer was able to take charge of more farmland. With increasing consumption and the advent of specialized labor, farmers required greater produce in exchange for more modern farming tools and other consumer goods. Specialization of labor, then, and emergent markets prompted the disintegration of rotation agriculture.

As farmers began to produce greater amounts of grain and produce, others began to specialize in nonagricultural trades. Increasingly compact communities became towns and then cities. Urban areas subsequently demanded supplies of farm produce and energy. Areas near urban centers were the first to not practice rotation agriculture. Ensuing intensive agriculture and animal husbandry became increasingly hard on the land. Fires to clear wasteland for cultivation, overgrazing, deforestation, and excessive irrigation each threw natural cycles of the various ecosystems askew.

Post–Industrial Revolution

The science and technology of Europe's Industrial Revolution further transformed humankind. At one time characterized as "passive adaptation," humans' interaction with the environment increasingly became "active." At present, the world's population is unprecedented. Average consumption levels continually increase as standards of living improve. Environmental impacts also increase in direct accordance with this growth.

Traditional agricultural districts (also suited for habitation) include plains in temperate areas, river valleys, and deltas. All are overcrowded. Vast areas of once-cultivated land are now covered with human habitation, factories, highways, mining operations, municipal facilities, and airports. Loss of cultivated areas has been quite high. Moreover, losses are commonly cumulative and irreversible. Farming districts compete to yield ever-increasing amounts of grain and a variety of produce. Once-natural biological boundaries are constricted and broken apart by cultivation. Humans scramble for land and rid reclaimed land of most of its living biota to increase personal wealth and prosperity. As a result, many biological resources, including wildlife, are becoming exhausted. Use of existing farmland has intensified and contributed to soil erosion and desertification.

Increasing pressure on limited resources such as fresh water, mineral deposits, and energy resources disturbs supply and demand equilibria. Not only do supplies contract, but demand increases exponentially with population growth and improved living standards. As once-dormant energy supplies are reintroduced into the energy flow of ecosystems, pollutant discharge also increases. Research, in fact, indicates that environmental impacts induced by population growth mainly occur in lesser developed countries (LDCs). Environmental impacts induced by increasing levels of consumption, on the other hand, mainly occur in developed countries. Specialization, commodity economies, and regional divisions of labor also contribute to environmental impacts.

Current Affairs

Population and environmental policies need to adapt to the social development of a country. In old China, laborers suffered from extreme poverty. With insufficient food, clothing, and housing, subsistence was a constant struggle. Since 1949, however, the state has paid more attention to the development of social welfare services. Housing conditions in urban and rural districts have improved. Social security, general levels of consumption, and environmental affairs have also gotten better.

Housing

From 1949 to 1986, investment in urban housing totaled RMB 255.6 billion yuan. Urban construction increased housing by 1.77 billion m^2. Investments from 1987 to 1988, by comparison, amounted to RMB 218.7 billion yuan and increased housing by 1.25 billion m^2. Per capita housing has also increased from 4.2 m^2 in 1979 to 8.8 m^2 in 1988.

Construction in rural areas has increased even more rapidly. From 1979 to 1989, the average annual investment in rural construction amounted to RMB 19.6 billion yuan. In 1990, rural housing covered 17.5 billion m^2, and per capita housing amounted to 16 m^2, a 142% increase since 1978.

Social Welfare

Nationwide health services are available now as well. By 1989, the number of medical establishments reached 207,000 with 2,568 hospital beds and 3,809,000 professional medical workers. Spread all about urban and rural districts, medical establishments are complemented by public health services run by collectives and individuals. The new social security system includes retirement expenses, free medical services, cooperative medical service, pension plans for the disabled and family members of deceased, social relief plans, labor insurance, and rural social security. The various welfare establishments include 37,000 homes for the aged with 761,000 beds. An estimated 580,000 persons have been admitted.

Moreover, social welfare establishments and employment opportunities accommodate 659,000 disabled persons. A 1987 survey estimates that China's disabled population is 51.64 million, or approximately the population of France. Hereditary and prenatal deformation accounts for the majority of handicaps. Disease incidence for mentally retarded children is 3–5%. In poorer and undeveloped areas, incidence of disease is higher at 10%. Approximately 50% of the 20 million mentally retarded children in China suffer from diseases due to social, cultural, educational, and genetic factors.

Consumption

Economic development has increased national income from RMB 26.03 billion yuan in 1952 to RMB 889.87 billion yuan in 1988. The annual average increase is 10.2%. Since 1979, the annual average increase has been 21%. Consumption of foodstuffs and material goods indicate the tremendous increase in living standards since 1978 (see Table 11.1).

Urban Development

Since 1949, plans for urban development have aimed to benefit economic development, improve public services, protect the environment, and enhance general living conditions. The availability of running water and effective water treatment facilities reveals progress in urban and rural town development. By the end of 1988, the daily use of tap water in urban areas reached 130 million tons, and utilization averaged 86%. Daily sewage water treatment is 1.98 million tons. Wastewater is treated 37.8% of the time. Urban central heating is found in 710 million m^2 of homes. The coverage rate of lawns in urban areas is 17%.

Table 11.1 Per Capita Foodstuff, 1978 and 1988 (in kg)

Year	Grain	Cooking oil	Meat	Poultry	Eggs	Seafood	Sugar
1978	195.46	1.60	8.42	0.44	1.97	3.50	3.42
1988	249.08	5.94	16.50	1.75	5.81	5.73	6.25

An estimated 165 million people, 47% of towns, and 8.5% of villages in China's more rural areas use running tap water. A total 624 million, or 71.8%, of the country's rural inhabitants have benefited from changes in the water supply. In 1980, 20% of the water in rural areas was safe for drinking, increasing to 66% by 1988. Also in 1988, electricity reached 96% of all rural towns and 63% of all villages.

Health

Improvements have additionally occurred in nutrition, medical services, and household and workplace sanitation. Before 1949, China's mortality rate was 25% and the average life span was 35 years. Since that time, mortality rates have improved dramatically. Mortality rates fell from 14% in the 1950s to 6.7% in 1987. The average life span has also increased nearly twofold to 67.3 years for males and 70.7 years for females.

National Income

Family-planning initiatives initially ran aground in the early to mid-1970s. Prior to effective population control measures, improved living standards simply fueled unabated population growth and environmental degradation. The expanding population soured relationships between humans and their surroundings. There were simply too many people competing for too few resources. What gains the Chinese economy had made since 1949 simply became diluted. National income in 1988 amounted to RMB 1.77 billion yuan, exceeding the 1952 national income by a factor of 21. If one accounts for population, the increase is still considerable, although less dramatic. Per capita income increased between 1952 and 1987 by a factor of 10.4. Aggregate consumption in 1988 amounted to RMB 797.1 million yuan, an increase from 1952 figures by a factor of 16.7. Excessive population growth, however, has tempered per capita consumption increases. Since 1952, per capita consumption has grown by a factor of only 8.7. If China's population had increased no higher than 700–800 million people, the state would have had ample funds to invest, increase production, and improve living conditions.

Culture and Education

The Chinese government has few funds to invest in culture and education because of the financial burden of population growth, resource depletion, and environmental degradation. Although the physique and general health of China's population is quite good, general education is less than optimal. Illiteracy rates have in fact dropped from 38.1% in 1964 to 20.6% in 1987, but the aggregate reduction is insignificant. Admission at colleges and universities is tremendously competitive with rates as low as 4.6%, compared to the world average university admission rate of 14.3%. Rates in Britain, France, and Japan are 23.4%, 18.3%, and 15.4%, respectively. An estimated 0.66% of China's population has a college education, 7% has a secondary school education, and 36.1% has at least a primary school education.

Urban Areas

Urban areas cover about 1% of the country's territory and accommodate 12% of the population. More than 30% of industrial enterprises and more than 20% of commercial enterprises are established in cities. Rural inhabitants migrate into the cities as reports of economic growth and opportunity spread. Populations in cities and towns increased from 19.4% in 1980 to 26.23% of China's population by 1990. The figures are estimated to increase to 31.5% by 1995 and 38.0% by 2000.

Towns symbolize civilization and progress. A natural tendency in social development, urbanization has brought about tremendous change to our surroundings. New problems that range from urban ecology to sociology have emerged. The advantages of cities, namely industrial productivity, are obvious. Data through 1984 indicate that average industrial productivity in 295 cities is RMB 11.9 yuan for every RMB 100 yuan of initial investment. This average figure is vastly exceeded in Shanghai (161.1 yuan), Ningbo (180.3 yuan), Bangfu (169.4 yuan), and Xiamen (137.1 yuan). In districts outside of the city centers, average productivity for each RMB 100 yuan of fixed investment is only RMB 75.4 yuan. The disparity stems from different productive capacities in and outside of the city proper.

Air Pollution

Drawbacks to urban life include overcrowding, housing shortages, traffic jams, and environmental degradation. Air, water, noise, and solid waste pollution are all quite serious. The proportion of energy in the atmosphere versus energy on the earth's surface determines climatic characteristics. For example, cities generally have more heat energy in the ambient air than do adjacent rural areas. Heat from household ovens and machinery emissions amasses in the sky. This residual energy is comparable to 30% of solar incidence. Because cities vary in traffic congestion, factory density, and weather conditions, the impact of residual heat varies considerably. Temperatures in cities, however, are generally higher than in surrounding areas. Cities of up to 10,000 people might experience temperatures 4°C higher than in the immediate surroundings. Cities that exceed 10 million might experience increases in temperatures as high as 10°C.

Hot air in cities gradually rises and is blocked by layers of city smog. Heat wells are formed and solar incidence exceeds reflection. This is commonly referred to as the greenhouse effect. When local wind speeds are below 3 m per second and temperature differences reach 5°C, a draft is formed that blows pollutants toward the city center. Because buildings in cities vary in size, height, and density, wind speeds and directions also vary. Average wind speeds for cities are commonly 20–30% lower than in surrounding areas. Depending on the region, air humidity is inversely proportional to air temperatures; therefore, in cities, where temperatures are high, air humidity may be low. Climatic variations in cities have a major impact on urban ecosystems. Humans, in particular, are easily affected by atmospheric pollutants. The *Report on the State of the Environment in China, 1990* indicates that atmospheric pollution is serious in large and medium cities. Total gas emissions, excluding township-run factories, amount to 8.5 trillion m^3, of which 14.95 million tons is sulfur dioxide, 13.24 million tons is soot, and 7.18 million tons is ash.

The average daily density of suspended particles nationwide in 1990 was 387 μg per m^3. In North China, suspended particles averaged 475 μg per m^3; in South China, the figure was comparatively low at 268 μg per m^3. Cities seriously polluted by suspended particles include Shijiazhuang, Nanchang, Jilin, Ulumuqi, Luoyang, and Tangshan. In spite of relatively good suspended particle averages in the South, sulfur dioxide pollutants are more serious there than in the North. Cities with serious SO_2 pollution include Chonging, Guiyang, Yibing, Nanchang, Shijiazhuang, Qingdao, and Ulumuqi. The average daily density of NO_2 nationwide is 42 μg per m^3. In North China, it is 47 μg per m^3; whereas, in South China it averages 38 μg per m^3. Acid rain is confined to certain areas but has intensified and spread over the last several years. The southern and southeastern portions of China are most seriously threatened.

Atmospheric pollution is a serious hazard to humans. It induces respiratory tract diseases and cancers. High incidence of cancer occurs in districts with a history of industry and atmospheric pollution. An investigation from 1976 to 1980 suggests that death rates in Tianjin from cancer increased by 20.6% over rates from 1973 to 1975. Death rates from lung cancer, in particular, increased by 50.4%. Districts in which lung cancer death rates exceed 30 per 10,000 are located primarily leeward from coal-consuming factories and near heavily congested areas.

Water Pollution

The main source of water pollutants in China is industrial wastewater. Ammonia, nitrogen, oxygen-consuming organic matter, heavy metals, arsenic, cyanide, petroleum products, and phenols are primary pollutants. Domestic sewage, mining, and cultivation also add water contaminants to China's supply. China's total wastewater emissions amounted to 35.4 billion tons in 1990. Industrial wastewater, excluding township enterprises, accounted for 70% of this total; domestic sewage accounted for the remainder.

An estimated 32% of China's industrial enterprises are located in 324 cities. Each of these cities suffers from serious water pollution. The Chinese *Bulletin on Environmental Conditions of 1992* indicates that pollution commonly originates near wastewater outlets along river banks. Industrial productivity, however, accounts for 70% of the national total. Hence, China's industrial sector would be crippled if excessive immediate regulation were imposed.

Groundwater quality in major cities was relatively clean in 1990. Spotty pollution points are more or less diluted. Main pollutants include excessive mineral salts, nitrate, sulfate, chloride, and phenols. Quality of groundwater is permissible in the following sampled cities: Xining, Lanzhou, Ji'nan, Qingdao, Nanjing, Hangzhou, Zhenzhou, Changsha, Nanning, Liuzhou, Guilin, Guangzhou, and Kunming. Groundwater is less than optimal in Harbin and Xi'an, among other places. Serious drops in water tables from excessive water use are found in 36 major cities including Shanghai, Tianjin, Suzhou, and Changzhou.

Water pollution is detrimental to health in several regards. Excessive nitrogen, phosphorus, potassium, sulfur, and varieties of their compounds ruin water quality, affecting aquaculture and general public health. Even tiny concentrations of mercury, cadmium, lead, chromium, or arsenic will poison animals, including humans. These minerals may enter human systems through food chains, accumulate as toxins, and cause cancerous mutations.

Noise Pollution

Noise pollution impairs mental and physical health. Noisy environments affect sleep, rest, and conversation. They induce auditory fatigue, vexation, excitation, poor concentration, insomnia, and speech impediments. According to the *Report on the State of the Environment in China, 1990*, urban noise sources include traffic (32.7%), household hoises (40.6%), and industrial or other noises (26.7%). Noise in China has unquestionably exceeded standards and continues to increase.

Solid Waste Disposal

Sources of solid waste in cities include industry, mining, household, and sewage. Collecting, storing, and disposing of each type of solid waste presents its own problems. According to the *Report on the State of the Environment in China, 1990*, industrial solid waste, excluding township-run enterprises, amounted to 580 million tons in 1990. An estimated 10 million tons was dumped into rivers. Legal dumping grounds are increasingly difficult to establish, however, because of the growing population density in cities. The economic costs of solid waste pollution amounted to RMB 7.95 billion yuan in 1986. As China's population and material consumption increase, waste management will become more important.

Solid waste contaminates soils. At present, only a few areas have been set aside for the disposal of nontoxic urban garbage. This is far short of requirements for industrial development. In addition to contaminating soils, solid waste also pollutes water via surface runoff, ground seepage, and wind. It further contaminates air with stenches from disposal and decomposition.

Rural Areas

Ecosystems in China's rural areas have undergone major changes since the advent of agriculture, forestry, animal husbandry, and aquaculture. Cultivated fields, tree plantations, pastures, and exotic plant species have encroached upon forests, grasslands, and natural vegetation. An estimated 800 million people, or 71% of China's population, live in rural areas. Their demands include increased agricultural production as well as land use for residential construction, factories, and highways. Land-use conflicts erupt when agricultural and nonagricultural needs compete. Population increases and improved living conditions further sharpen these conflicts.

Intensifying land use is one method of alleviating population pressure. Researchers claim that China's potential arable land base is limited. Hence, land conversion will not be an option much longer. Increasing grain and produce outputs on currently cultivated areas, however, remains a viable option. From 1975 to 1982, grain output increased by a factor of 31 with added investments in fertilizers, pesticides, and energy availability. Investments increased by factors of 37.5, 2.6, and 11, respectively.

Increases in population and production in China's rural areas disturb ecosystems in several ways. Soil erosion, toxic pollutants from use of chemicals in agriculture, and industrial pollution are but a few of the most obvious damaging factors.

Soil Erosion

Cultivated areas cover less than 100 million ha, or less than one-tenth ha per capita. This figure includes 6.7 million ha of saline-alkali land, 4.0 million ha of wetland areas, 36.7 million ha of sloped terrain, and 20.0 million ha of polluted farmland. Net loss of farmland is tremendous. Railways, water projects, and mineral and energy resource mining have typically out-competed reclamation for farmland. From 1957 to 1980, the annual net loss of cultivated land averaged 545,000 ha. During the Sixth Five-Year Plan, the aggregate loss of farmland was 2.46 million ha, an annual average of 522,000 ha. Then, from 1985 to 1986, net loss of cultivated areas amounted to 1.65 million ha. Township enterprises have contributed to these net losses in cultivated areas. For instance, in Jiangsu Province, an estimated 58% of the 13,400 ha of lost cultivated areas in 1984 went toward township-run enterprises. Annual net loss in cultivated areas is 8%.

Land is typically underused. Approximately 30.4% of farmland is low yield, and only 20.8% is considered high yield. Sterile soils have worsened overall productivity. This factor alone hinders the future of agriculture in China and threatens general stability.

Toxins

An additional threat to the long-term productivity of soils and water supplies is the use of chemicals in agriculture and resource management. By the end of 1983, the accumulated output of pesticides totaled 8.15 million tons, or 2.10 kg per ha. Pesticides contain chlorine, which pollutes the environment, farm produce, and livestock products. The State Council has taken action to ban the use of pesticides such as BHC and DDT. Long-term environmental impacts from the use of organic chloride in pesticides will continue through the year 2000.

Nitrogen fertilizers are very popular in China. Although production in 1980 amounted to 9.43 million tons, the utilization rate was only 40%. Fertilizers containing phosphorus in the same year totaled more than 3 million tons. Runoff after application averages 1%. Environmental damage results when excessive concentrations of nitrogen and phosphorus contaminate rivers and lakes. Economic development has permitted more affluent peasants to use chemical fertilizer in the place of manure. The utilization rate of excrement for manure is now only 50%. This leaves the remaining 50% as a new source of solid waste to manage.

Industrial Pollutants

Industrial waste is another primary source of solid waste that is detrimental to rural ecosystems. In the early 1980s, sulfur dioxide from industry in urban districts polluted 540,000 ha of farmland. Polluted water contaminated an estimated 378,000 ha of farmland, of which 25 ha was seriously impacted. About one-third of urban garbage and excrement is transported to rural areas for dumping and use as fertilizer. The residual pollutes the air and water of suburban areas.

Economic losses of unmanaged solid waste pollution are tremendous. Grain output fell by an estimated 16 billion kg in 1980 due to pollution. Solid waste and pesticides impacted an additional 41.4 billion kg. Economic losses amounted to RMB 23–26 billion yuan. European and U.S. markets refused to accept many of China's farm and livestock products because of the pollutants.

Township Enterprises

China's surplus labor amounts to 200 million. Growth in agriculture, forestry, animal husbandry, and fishery industries will help to alleviate problems in rural areas, but urban areas need to grapple with the remainder. From 1949 to 1983, China's annual population growth was 1.9%, whereas urban growth rates were 4.3%. An estimated 70 million people, or 20% of the annual increase in rural inhabitants, move to the cities from rural areas. This amount of people far exceeds demand for labor in urban areas.

The Third Plenary Session of the Eleventh Central Committee of the Chinese Communist Party established guidelines to rejuvenate and open the economy to the outside world. A central component of the guidelines, township enterprises, subsequently flourished. Output from township enterprises accounts for 50.4% of the total output from China's rural economy. The enterprises have become a powerful force in the national economy. In 1983, an estimated 18 million township enterprises produced RMB 610 billion yuan worth of merchandise, or 28% of the nation's industrial output. Scattered about some 2,000 counties, they have tremendously shifted the distribution of industry and alleviated problems of surplus labor and urban migration. In 1986, township enterprises employed 80 million people. Rural economies have unquestionably benefited from the initiative.

In spite of the benefits, these enterprises are characterized by poor distribution, outmoded technology, obsolete equipment, poor management, and heavy consumption of resources. Most lack pollution prevention and treatment facilities. In 1987 gas emissions from township enterprises amounted to 300 million m^3, or 32.7% of the national total. Polluted water in the same year amounted to 4.9 billion tons, or 13.2% of the national total. Pollution from township enterprises accounts for more than 15% of China's total. Since the vast majority of enterprises tend to be near coastal areas, pollutants commonly enter agriculture's nutrient cycling and inflict damage on humans and their surrounding ecosystem.

Electroplating, paper, and printing industries are a few of the heavily polluting industries that contaminate the surrounding environment. A single papermaking factory may pollute a river and impair the daily productivity of entire villages downstream. The impacts are far-reaching and include general environmental degradation, reduced agricultural activity, high disease and death rates, and reduced standards of living.

In the year 2000, township enterprises will offer employment to 170–200 million currently unemployed laborers, or 40% of the total rural labor. Productivity should reach RMB 1.36–1.50 trillion yuan, or 75% of rural areas' economic output. Some optimists claim that these figures are conservative and that actual productivity may be valued at RMB 1.70 trillion yuan in 1995 and RMB 5.50 trillion yuan in 2000. In spite of their increasing appeal, township enterprises are prone to become heavy polluters. Leadership at various government levels should steer the course of events away from becoming heavy pollution sources. Township enterprises need to abide by strict pollution guidelines from the very beginning.

Population Characteristics

Certain characteristics of China's population are degenerating as the population increases and the environment deteriorates. Although genetics and mutation play decisive roles in determining the physical and mental characteristics of people, environmental surroundings unquestionably have an impact as well. Throughout history, humankind has survived through adaptation and natural selection. Humans have thrived where natural surroundings permit. Where surroundings are less than optimal, humans have used both mental and physical capabilities to manipulate their environment. A brief review of the cycle of life will more clearly reveal how genetics and the environment in accordance with natural selection help human civilizations to progress.

The quality of an individual's gametes depends exclusively on the genetic source and an individual's environment during puberty. Once gametes meet, the growth stage of an embryo is perhaps the single most vulnerable stage of an individual's existence. A pregnant woman under physical and mental stress may give rise to a malformed fetus. Polluted surroundings will further impact fetal formation. For example, radiation can interfere with nerve fibers and the formation of optic senses. Radium therapy may also induce cleft spine, skull, and skeletal formation.

Postnatal developments refer to the growth and development of physical and intellectual abilities. In spite of hereditary disease, these developments are again largely determined by an individual's environment. Food availability is, of course, a preeminent variable of one's surroundings; it has traditionally been a problem in some areas of China. In spite of scientific research indicating that nutritional requirements include a minimum 1,000 kg of grain per capita per year, China's per capita availability remained at 300 kg per capita per year for years. It was not until 1986 that per capita grain output reached 400 kg. Living standards have greatly improved, but nutritional requirements are still not being met. Per capita consumption of protein at 67 grams per day has just reached the lower limits of recommended intake. In spite of recent bumper harvests, per capita grain outputs have not increased. According to the State Physical Culture and Sports Commission, the State Education Commission, and the Ministry of Health, physical development of children and adolescents has taken a turn for the better. Disparities relative to more developed countries still exist, however.

Pollution threatens the health and livelihood of people. In heavily polluted areas, people are apt to be irritable, anxious, or distraught. In quiet, clean, and more aesthetic environments, people are generally more relaxed and cheerful. This sort of surrounding will help alleviate fatigue, nervous tension, and depression. Mental health further influences physical health and one's immunity to disease.

According to Russian experts on longevity, 15.2% of an individual's longevity depends on genetics, 20–30% depends on one's surroundings, 40–50% depends on one's manner of life, and 10% depends on available medical services. During puberty, one's way of life has a tremendous impact on both health and longevity. Eating habits, addictions, clothing, housing, and means of transport each play a role in this regard. Eating habits may, in fact, be the most influential variable. Nutritional levels of food and the degree of toxicity in foods have a major impact on one's health. Excessive drinking or smoking are widely acknowledged health risks that may reduce longevity. Meeting basic necessities of life, sanitation, and medical care are each exceptionally important in preventing and treating disease.

Humans and their environment clearly influence one another. As stated, an individual's surrounding will determine many of the individual's physical and mental characteristics. By the same token, individuals carry tremendous power and influence in shaping wildlife habitats and entire ecosystems. Science and technology aim to enhance human capabilities and our efficient use of natural resources. We have the option to use the technology to either benefit or further degrade the environment. If population characteristics improve and public awareness campaigns are successful, people will increasingly realize the value of adhering to natural laws of ecosystems. People will actively improve and protect the environment from further degradation.

The environment has become a preeminent global concern. In recent years, state policies have achieved positive results in environmental protection. Public concern for the environment remains, however, quite weak. Traditional notions of the environment and an exclusive concern for the economy persist among some industrialists. These people seek immediate and high levels of productivity without regard for the environment. Poor awareness more than anything else contributes to people establishing inefficient, heavily polluting township enterprises. In some cities, polluting industries have simply been relocated to the countryside where people are less aware of pollution problems and authorities have limited knowledge or capacity to act.

Environmental pollution is a prominent factor affecting rural peasants' quality of life, according to areas that have undergone rapid development, such as Jiangsu Province. Reverence for small-scale production and inefficient, backward methods of production hinders efforts to protect the environment in more remote areas. Educating resource managers about the essentials of ecology and environmental protection is, thus, vitally important.

Current levels of education, science, and technology in China are inadequate for the country's development needs. A national survey of 1% of the country concluded that there were 180 million illiterate and semi-illiterate people in China. An estimated 57% of the total population in China have education below the primary school level. This segment of the population has become an inert force in China's progress. Manpower without limited amounts of education can go only so far. China currently has an oversupply of people in low-level unskilled jobs. People qualified for high-level, more skilled jobs are in demand. The economy will stagnate until these positions are filled with competent personnel. Social, economic, and environmental objectives will all fall short until China produces enough educated personnel to fill the necessary roles.

Use of natural resources falls far below that in other countries. The trend will continue until China educates competent personnel to increase utilization rates of its resources. Coal consumption, for example, per US$10,000 of GNP was 21.11 tons in 1980. Japan and Germany, by comparison, consumed 4.35 and 4.86 tons, respectively. Average utilization of energy resources is 30% in China. Utilization rates in Japan, the United States, and other European countries are 57%, 51%, and 42%, respectively.

Unemployment will also remain a primary concern for China in the years ahead. An estimated 21 million youth have recently entered the job market. In the next decade, the number will decrease but remain substantial at 15 million. China will unquestionably find it increasingly difficult to employ such large numbers. To guarantee social stability, state industry and agriculture will be forced to hire unnecessary personnel. Current surplus labor in urban areas amounts to 20 million people; surplus labor in rural areas amounts to 200 million. Hiring unnecessary labor in fact hampers labor productivity and efficiency.

By the year 2000, pollution is bound to worsen with a doubling of the GNP and industrial development. Urbanization also will increase to the point where 30% of the population, or 400 million people, will inhabit cities. Environmental protection, then, is as important in urban areas as it is in the rural countryside. If urban areas deteriorate, productivity will fall along with improved living standards. The same holds true for rural areas. Population growth and development will damage rural ecosystems because of increased use of pesticides and fertilizers. Waters will be increasingly contaminated as general productivity and living standards fall.

Coordinating Population Policies and Environmental Affairs

Circumstances in urban and rural areas are so unique that blanket policies will not work. Policymakers need to discriminate between these two domains. Urban environments will be more keenly impacted by economic expansion over the next decade. Although rural environments are of equal and perhaps greater magnitude, the country must focus on the greatest problem at hand. At present, these problems occur in urban areas.

Urbanization will soon curtail because of tradition and current economic affairs. Optimal conditions in an urban environment include first-rate public facilities, a streamlined administration, sound financial management, and a clean environment. Public facilities refer, in this case, to production, livelihood, and cultural facilities. A modern city will generally have adequate water supplies; water treatment facilities; a network of roads, bridges, and tunnels; public transportation; sufficient power supplies; postal services; urban parks and trees; and a generally clean environment.

These components should be the skeletal framework of urban design. Urban areas might further be defined into factory, cultural, commercial, and residential quarters. A U.S. expert in management explains that "management is an internal organ of the community, its raison d'être lies in its efficacy." Efficient management and administration are primary concerns for urban development. Organized procedures and a democratic method are two essential components of efficiency. Administrative responsibilities include industrial production, financial affairs, market activities, communications, aesthetics, and environmental protection. In simple form, administration really grapples with human interaction. It steers relations between productive forces and the community.

Urban areas should strive to optimize profits in terms of the economy, the society, and the environment. Economic profits refer to optimizing productive potential. Societal profits refer to optimizing living standards. Environmental profits refer to optimizing a city's natural and artificial surroundings for present and future generations.

Objectives for environmental protection in China are to bring pollution under control, enhance urban environments, and alleviate pressures on rural ecosystems by the end of the century. To achieve these ends, environmental policies need to be intimately tied to China's other foremost concerns—population control and socioeconomic development. The following four measures may help to achieve these objectives: (1) implement strict urban development guidelines, (2) enhance urban environments by preventing the "four hazards," (3) enhance rural ecosystems according to regional conditions, and (4) implement family-planning and public awareness initiatives.

Urban Development Guidelines

Strict guidelines for urban development aim to coordinate growth with environmental protection. Preventative measures are typically the most cost-effective approach to a clean environment. The country needs to categorize current conditions before establishing guidelines. Assessing city populations is an important first step. Cities with a nonagricultural population of more than 1 million are classified as "extraordinarily large cities"; those with a population between 500,000 and 1 million are considered "large cities"; those with a population between 200,000 and 500,000 are considered "medium-sized cities"; and those with a population of less than 200,000 are considered "small cities."

The State Council has formulated guidelines to control growth in large cities, to refine growth in medium-sized cities, and to encourage growth in small cities. This strategy will help to more evenly distribute growth and its unfortunate side effects. Large cities play an important role in developing the national economy. They also contribute to a plethora of problems including housing shortages, traffic congestion, sewage blocks, degenerating public facilities, and environmental pollution of all sorts. Each of these problems hampers industrial productivity and the daily lives of urban inhabitants. To solve their many problems, cities should at once be developed and controlled. This is to say that cities should be encouraged to develop as pivotal centers of socioeconomic development. Population growth and all of its associated problems, however, must be kept in check.

Medium-sized cities produce profits larger than small cities but smaller than large cities. They generally maintain most of the aforementioned public facilities, but do not have the infrastructure required for complete efficiency. Refining facilities in medium-sized cities is, therefore, the main objective for this type of municipality.

Encouraging growth in small cities distinguishes China's development strategy from that of other countries. As socioeconomic centers of their region, small cities stand to benefit enormously as China shifts gears from a predominantly agriculture-based economy to a more diversified industrial economy. Small cities are also crucial to settling China's unemployed, "floating" populace. Currently immune from the "municipal diseases" that afflict larger cities, the development of small cities and towns is a primary objective of China's development scheme.

Preventing the "Four Hazards"

The four hazards include air, water, noise, and solid waste pollutants. (See Chapter 9 for a discussion of measures to prevent water pollution.) Measures to prevent air, noise, and solid waste pollutants are as follows. Coal washing should be raised to 36% of total consumption by the year 2000. Sulfur fixation will also help to reduce emissions of sulfur dioxide, the leading cause of acid rain. Adopting dust removers may additionally help to limit dust emissions. In cities where small coal burners are common, coal washing and shaping will be highly effective. Stack desulfurization and dehydrosulfidation should also be implemented. Tree planting is an additional effective means to control dust in the ambient air.

Laws and regulations, efficient management and administration, science and technology, and public awareness campaigns will each help to prevent noise pollution. Preventing excessive solid waste calls for strengthened leadership, innovative ideas on multiple use, implementation and enforcement of effective regulations, and administrative efficiency. China needs to improve and renew its technologies; shift to cleaner, more plentiful energy supplies; and reduce waste emissions. Promoting technologies that reuse waste is an important measure that would meet all three objectives. Research and case studies should be encouraged as well.

Enhancing Rural Ecosystems

By the year 2000, the rural population will reach 900 million. Although environmental protection in urban areas may be a priority at present, rural areas command attention because of their size, their unique attributes, and their importance to the livelihood of all Chinese citizens. Any and all policies should be conditional to development research and China's objective to encourage growth in small towns. Township enterprises should be encouraged where they suit local conditions. Inefficient, heavily polluting enterprises need to be reformed. Use of organic fertilizers should be encouraged over use of chemicals. Also, use of rural energy sources including natural gas, hydroelectric power plants, fuel forests, and small coal pits should be encouraged to ease pressures on grasslands, production forests, and wildlife preserves. Solar, wind, geothermal, and other new energy sources should also be exploited. Biological control and limited use of low-toxicity pesticides may help to prevent natural disturbances where not desired. New methods of management, science, and technology will also help environmental protection efforts.

Agriculture is the foundation of China's economy. Increases in grain, cotton, and other farm products are of primary importance to the stability of the whole country. The government should in no event cut back on agricultural productivity. Approximately 92% of China's farmland is located in overpopulated areas in the Southeast. Both per capita cultivated areas and per capita areas in reserve are very low. Local officials are therefore pressed to open up mountainous areas, beaches, grasslands, and inland water areas to cultivation. Alternative or at least complementary measures might include organic agriculture, agroforestry, forestry, livestock production, and fisheries.

Increasing Public Awareness

As the population increases and people's appetite for better living conditions grows stronger, China has few options but to keep population growth in check. An additional effective measure is to educate the existing populace of reasons for family planning and natural resource management. If general awareness increases, efforts to expand the economy and protect the environment will be made simple.

Enhancing current population characteristics is made difficult by poor communications infrastructure and ties to traditional methods. If education, eugenics, and healthy life-styles can be continued and promoted among families, China will have taken a giant first step. Population, economic, and environmental policies need at this point to be intimately related and sensitive to local circumstances.

If successful, China will have transformed a once-vicious cycle into a more constructive one. The current cycle appears as follows: population expansion → strengthened economic activity → environmental degradation → poor population characteristics/awareness → continued population expansion. The new cycle may appear more like the following: moderate population increase → sound economic development → sound environmental protection → enhanced population characteristics/awareness → effective control of population growth.

12

Population Levels and Environmental Affairs: A Comprehensive Analysis

The relationships between China's population and its forest, grassland, mineral, water, and energy resources have thus far been discussed in detail. China's population clearly impacts each component of its vast resource base. Conversely, the country's natural resources have impacted population growth and distribution. In this chapter, we will further analyze the relationships between China's population and its natural resources.

Mechanics of the Relationship

Science and technology have amplified human impacts on the environment. Human impacts have had more influence in some areas of the country, however, than they have in others. China's social economy is, hence, unbalanced and characterized by a "dual structure." Modern urban areas with tremendous productive potential contrast with the vast countryside's lower productivity. Urban enterprises with advanced technology and efficient management contrast with rural industries' backward technology and poor efficiency. International theories on efficient management contrast with traditional, practiced methods. The resulting unequal economic development has created a business environment unique to China.

Excessive Use

China's increasing population and the growing demand for improved living conditions have placed tremendous pressures on the country's current resource base. Increases in rural population, in particular, impact forest, grassland, mineral, and water resources. Although in many cases renewable, these resources are extracted in some areas at unsustainable rates. Cultivated areas, for example, are best managed by "planting one year and leaving fallow the next." This scenario allows for perpetual use. In China, however, the heavy demand for grain and other products precludes such methods of production.

In spite of an abundant total resource base, China's per capita resource availability is rather limited. Population pressures exerted on the natural resource base exceed similar pressures in other countries. China's unique natural ecosystems, hence, suffer serious degradation on both an aggregate scale and a per capita scale relative to other areas of the world. Animal husbandry and agriculture threaten grassland ecosystems, for instance, in arid and semiarid northwestern regions of China. Animal husbandry has additionally increased demand for additional foodstuff. Successful breeding inevitably stretches herds beyond the land's carrying capacity, at which point overgrazing and packed soils degrade pastures.

Land reclamation for agriculture additionally destroys large areas of unique grassland vegetation. Geological condition, soil quality, and windy, arid climates often increase the likelihood of desertification. On China's Loess Plateau, local peasants denude forests for fuel wood. Moreover, subsequent cultivation intensifies soil erosion and decreases soil fertility. Even if peasants are aware of the natural processes of nature, poverty and despair leave them with few options.

Slash-and-burn agriculture also afflicts unique areas in mountainous regions of Yunnan, Guizhou, and Sichuan provinces. In these poor and backward areas, high birthrates further exasperate environmental damage. These areas have, hence, fallen into a vicious cycle of population increase → environmental degradation → impoverished economy → poor public awareness → population increase.

In recent years, China's leaders have become aware of the need to control population growth and protect the environment. They now realize that short-term alleviation of poverty can cause long-term environmental catastrophe and even wider-scale poverty. In spite of their increased awareness, China's legislators have a long, arduous task in front of them in establishing, implementing, and enforcing the new policies.

Labor Supply

Rapid population growth in rural areas expands the labor supply beyond existing demand. The alienated and unemployed rural populace then moves into urban areas and further disrupts urban employment patterns. Township enterprises have helped to stem the tide of unemployed labor. These enterprises have played an important role in providing employment to rural surplus labor, which has helped to alleviate the rapid rates of migration to already saturated urban areas. Furthermore, these enterprises have helped to breathe new life into stagnant rural economies.

Many of these township enterprises, however, carelessly pollute the environment. Short of funds, the enterprises rely on backward machinery, which emits pollutants into rivers, fields, and the atmosphere. Smelting is a principal polluting industry. Concentrations of sulfur dioxide exceed state standards by enormous amounts for several square kilometers surrounding an operation. In many regions near smelting operations, hillsides and cultivated areas are barren and sterile due to acid rain. Cultivated areas often do not return to normal productivity even after leaving fields fallow for 10 or more years. Arsenic refining is common in Hunan, Guangdong, Guangxi, and other provinces. White arsenic, As_2O_3, is especially toxic. Wherever poisonous gas residues drift, trees wither, soils deteriorate, and ecosystems generally deteriorate. Employees frequently suffer from acute poisoning after only short periods of exposure. Limbs and occasionally entire bodies can become infected. The poison is a serious carcinogen.

Paper-producing township enterprises play an important role in overcoming paper shortages. Rural paper mills, however, are small in scale and scattered, with backward technology and crude equipment. Waste emissions are tremendous. Variability exists, however, between the township-run mills and state-run mills. The amount of wastewater per 1 ton of paper in township mills is double that in state mills. Moreover, the pollution in township mills' wastewater has a concentration 3 times that in state mills. Suspended particles are also higher by a factor of 13. A small-scale township enterprise might easily pollute an entire river, reservoir, beach, or farming area.

In 1987, township enterprises were responsible for 32% of the country's gas emissions, 13.2% of the country's wastewater, and 15% of the country's solid waste. Waste from township enterprises is rarely treated. It inflicts tremendous damage to rural environments and general public health.

Urban Infrastructure

Natural population growth in urban areas, in addition to migration from rural areas, tests the capacities of urban infrastructure. In the past few years, construction in China has accelerated to combat the trend. Regardless of government investments in urban infrastructure, the efforts still fall short of new demand. Many large cities are troubled by water and energy shortages. Traffic congestion is also a problem; road area per capita in 14 provinces and cities directly under the jurisdiction of the central government is less than 3 m^2. China's cities have far less to offer by way of infrastructure than do other major cities of the world (see Tables 12.1 and 12.2). Many factors, of course, help to explain this; the most prominent, however, is China's excessive population growth.

Insufficient facilities in China inhibit construction and growth. Limited government funding further prevents effective environmental protection. Economic development will gradually help to fund new facilities and protection efforts. Population growth and urbanization, however, are increasing at rates faster than economic development.

Table 12.1 Public Facilities in Selected Cities

Public facilities	Budapest	Moscow	Tokyo	London	Paris
Area of roads per capita	9.5	7.2	9.68	26.3	9.3
Number of public traffic vehicles per 10,000 persons	15.31	13.53	—	30.7	—
Daily water supply per capita (liters)	320	520	400	295	320
Wastewater treatment rate (%)	39	95	90	95	93
Rate of domestic gas burners (%)	71.5	98	85	100	—
Public lawn per capita (m^2)	7.74	18.8	1.6	30.4	12.21

Economic Development

Since 1949, China's economy has undergone considerable growth. GNP is high relative to world averages. Rates of increase in per capita national income, however, are far lower than increases in total GNP (see Table 12.3).

Past relations between population growth, production, consumption, industry, and employment structures suggest that massive increases in population spell disaster for developing economies. Population growth will cause reductions in per capita income, drops in general consumption, reduced production, and contraction. Downscaling payrolls will make it more difficult for individuals to support their families, even though per capita equipment rates and labor productivity are down. Low-level production, then, makes increases in per capita income unlikely. Professions in nonfoodstuff businesses have little opportunity to grow. Moreover, limited funding restrains development of culture, science, and education. People are instead subject to low levels of education. They insist that additional sons bring happiness and security in old age.

Excessive population growth and backward economies are, therefore, complementary. Backward economies, furthermore, cannot afford large-scale environmental protection. Only in the past several years has the government placed environmental protection on the list of policy debates. China's resolve to remedy environmental pollution is now limited only by financial needs.

Our analysis suggests that environmental impacts from population growth are characterized more by a network of events than by a simple cause and effect chain of events. Population increases, increasing demand on various resources, resource degradation, urbanization, unemployment, inefficiencies in the economy—all feed off one another. Solutions, therefore, should not exclusively stem from either population control or environmental protection. The interrelationships need to be carefully examined. Population levels, resource availability, economic development, and environmental quality are each an essential component of a sound, sustainable development path.

Table 12.2 Public Facilities in China's Major Cities

Public facilities	Beijing	Tianjin	Shenyang	Shanghai	Nanjing	Qingdao	Chungqing	Xi'an	Ningbo	Wuhan
Rate of water supply (%)	87.4	100.0	86.8	100.0	99.2	97.0	90.8	94.5	96.7	93.2
Number of public traffic vehicles per 10,000 persons	8.9	4.2	3.6	8.6	5.5	5.9	5.8	4.3	3.3	5.9
Rate of domestic gas burners (%)	85.5	82.6	65.9	58.8	60.9	47.3	47.9	28.5	36.5	30.0
Public lawn per capita (m^2)	6.1	2.1	7.6	1.2	6.9	3.5	0.9	1.5	1.2	2.4

Table 12.3 China's GNP and National Income, 1952–1990

Year	GNP (billion yuan)	National income (billion yuan)	National income per capita (yuan)
1952	—	58.9	104
1975	358.8	301	315
1980	447	368	376
1985	855.7	702	672
1988	1,398.4	1,173	1,078
1989	1,578.9	1,312	1,189
1990	1,740	1,430	1,251

Figure 12.1 Population Distribution in Urban Centers

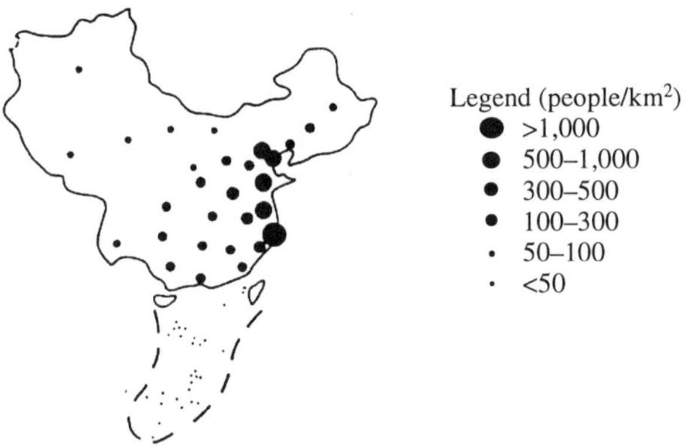

Anthropogenic Impacts

The analysis thus far has presumed that the size of China's population is the most important variable in determining the state of China's environment. Although this is perhaps true, multiple other characteristics of China's population come into play: density, age structure, scientific and cultural traits, and so on.

Population Density

Discussion on the impact of population size on the environment only infers the effects of increases in density. China's population is unevenly distributed (Figure 12.1). The Southeast comprises 43% of the nation's territory and bears 91% of the country's population. The Northwest comprises 57% of the nation's territory but bears a mere 9% of the population. Water, gas, and solid waste discharge in the East forms a large proportion of the country's waste emissions (Table 12.4). The distribution of pollutant and population density indicates that

Table 12.4 Waste Emissions in West and East China, 1989

	West China		East China	
Waste emission source	Emission	Rate (%)	Emission	Rate (%)
Industrial wastewater (million tons)	930	3.7	24,270	9.63
Waste gas (billion m^3)	630.2	7.6	7,676	92.4
Sulfur dioxide (million tons)	1.129	7.2	14.51	92.8
Soot (million tons)	1.171	8.4	12.809	91.6
Industrial solid waste (million tons)	7.88	15.0	44.77	85.0

Figure 12.2 Sulfur Dioxide Emissions Per Unit Area in China

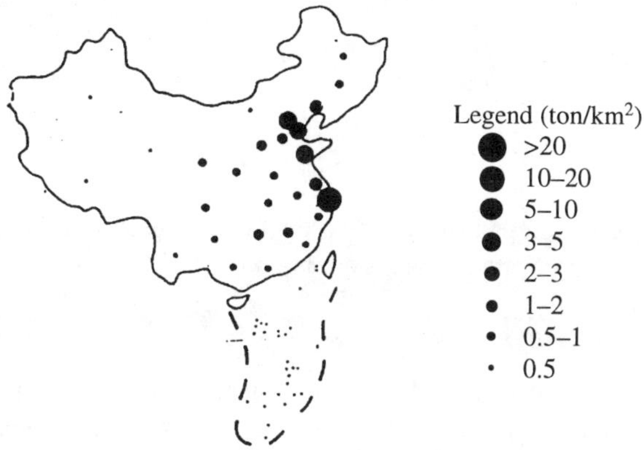

there is a direct correspondence under certain conditions (compare Figure 12.2 with Figure 12.1). The Southeast is generally endowed with greater natural resources. It has a long history of resource use and economic development.

Demography

A large and steadily growing number of young men and women at childbirth age characterize China's present age structure. Growth trends suggest that childbirths will climax in the year 2015. It is hoped that birthrates will gradually fall from 2.4 to 2.2 and on further to 2.0. Population sizes at these various rates are shown in Figure 12.3. Even if strict childbirth policies are enforced, growth rates will not fall until the beginning of the next century. Pressures from overpopulation will unquestionably continue for the next 50 years or more.

Natural resource availability will similarly not improve for some time in spite of current efforts. Water resources should more or less remain static, and a steady decline in cultivated land is probable in the foreseeable future. There will be few increases in forests and grasslands in the immediate future. However,

Figure 12.3 Population Growth Trends According to Birthrates

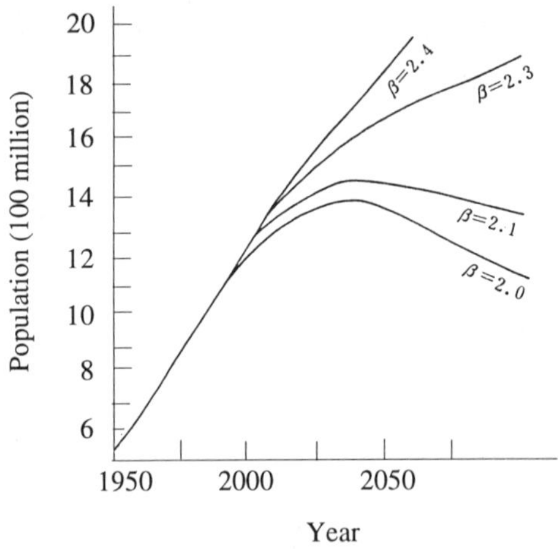

optimists also do not predict substantial reductions in resources throughout the century—though per capita resources are bound to fall.

Environmental Protection Measures

Intellect sets humans apart from other animals. It allows humans to quantify and examine natural processes, as well as their impact on the environment. This knowledge, in turn, widens the range of options for use of resources. In the past, low levels of science and technology precluded humankind from having a major impact on the environment. As science and technology developed, however, humans became more powerful relative to their surroundings. Their ability to manipulate and transform natural resources has increased tremendously. Inadequate knowledge of ecological principles, however, has upset China's many unique ecosystems. Increasing public awareness will help to integrate environmental protection with economic expansion. Science and technology, then, can be at once dangerous and invaluable to the environment, depending on respect for ecological principles.

Public awareness of ecological principles shapes human impact on the environment. Low levels of awareness are generally cause for intensive extraction of resources and myopia; high levels of awareness are generally cause for more sustainable extraction rates and long-term planning. During China's Great Leap Forward, the country focused all efforts on increasing steel production. An economic miracle was to be attempted at the expense of vast forest resources. In less than two years, efforts waned, but once-expansive areas of forests were denuded. Climatic changes and soil erosion include two of the continuing impacts of the disaster.

Once knowledge of science and technology has been achieved, actual prevention and treatment require manifold efforts that include the following: (1) enhancing production methods of industry, (2) instilling waste treatment and energy-saving technology, and (3) reducing pollutant discharge. The country needs to simultaneously maximize use of available resources and address waste management. Economic growth and public awareness will guarantee a supply of qualified scientists and technicians and a competitive industrial sector that supplies environmental protection technology.

Carrying Capacity of the Environment

Carrying capacity refers to the earth's capacity to sustain a population. It is the maximum population size that can be kept alive for any given ecosystem. International population ecologists have defined the world's population carrying capacity as follows: "the size of population that can be supported by the world's resources in a long-term stable state under conditions that the biosphere is kept intact and that non-renewable resources are not exhausted." The earth's carrying capacity is constrained by many factors and may hardly ever be reached.

Determining the earth's carrying capacity is an intense issue. Productivity of the land is without question limited. It follows that the potential number of inhabitants on the face of the earth is finite. According to research, an average individual requires 2,400 kC of energy per day to maintain health, or 876,000 kC per year. Plants' biomass production is 6.4×10^{17} kC, but only 1% of plants are edible in one form or another. Therefore, the earth might sustain a population of 8 billion people.

Scientists have long studied China's appropriate carrying capacity. In 1957, Sun Ben-wen suggested that in terms of grain output and employment, China's appropriate population should be no larger than 800 million. In 1980, Tian Xueyuan and Cheng Yuguang maintained that in terms of general laborers, the appropriate population in 100 years should be no larger than 650–700 million. Hu Baosheng and others estimate that on the basis of some 20 different criteria, China's population should be kept between 700 million and 1 billion. Song Jian and others accounted for foodstuff and freshwater resources, and estimated that the country's population should be kept under 700 million in 100 years.

Scientists generally concur that natural resources and the environment are primary restraints of the population carrying capacity. Surveys of China's current situation suggest that whereas the environment has improved locally, it has deteriorated on the whole. Prospects are worrisome. From an environmentalist's viewpoint, China has far exceeded its appropriate population capacity. Anthropogenic environmental disasters of late attest to this view.

In Guizhou Province, Southwest China, for example, per capita land resources amount to 0.55 ha. If one were to account for the province's rugged terrain, per capita farmland amounts to 0.06 ha. In addition, approximately 80%

of the farmland is plagued by infertility to some degree. Excessive use continues to erode cultivated areas.

According to international standards, 300 kg of grain per capita per year is the suggested minimum for maintaining social stability. Since 1949, per capita grain output in Guizhou has fallen below this suggested minimum every year except 1957 (with a harvest of 304 kg per capita). Even if the province were able to produce 300 kg of grain per capita per year, its population still exceeds recommended levels at 25 million people. Needless to say, the land is currently overworked.

Deep groundwater reserves and terraced hillsides characterize the area's hydrologic cycle. At present, a shortage of drinking water affects more than 5 million people and 3 million livestock. Nearly 10 million people are without clean drinking water. Guizhou Province is also short of petroleum and natural gas. Coal deposits and the potential for hydroelectricity are, however, abundant. Energy resource use is made difficult by poor infrastructure, specifically transportation. Per capita energy consumption is, hence, lower than the national average. Firewood accounts for 76% of the rural area's energy consumption. Timber harvests, therefore, exceed annual growth by 3 million m^3. Research indicates that if forest coverage drops below 20%, lumber supplies are no longer self-sustainable. Maintenance of natural ecosystems requires a minimum 30% forest coverage. In terms of its forest resources, the population in Guizhou should not exceed 5 million people.

The analysis of Guizhou indicates that the province's land, water, and energy resources are currently overexploited. This is also the case with many other provinces, autonomous regions, and cities in China. As China's population continues to increase, environmental impacts will continue to have a devastating influence on the economy.

A Role for China's Human Resources

China's population problem is a central component of any social, economic, or political policy. The problem emerged due to certain social and economic factors, combined with gross policy errors. Retrospection will help to avoid similar problems in the future. The population is an asset as well as a liability, however, when it comes to implementing environmental protection. China has the potential to transform its rich manpower resources into an active and efficient movement for environmental protection.

Currently, basic means of livelihood such as food, clothing, housing, employment, education, transportation, and medical care are at jeopardy. An aging population and higher childbirth rates will present another gamut of problems in the years ahead. Our abilities at harnessing nature's productive potential are limited only by ignorance of natural ecosystem laws. If we use our entire capacity, humankind will overflow with consumer goods. A common saying runs as follows: "A man has not only a mouth, but two hands." In other words, humans need to consume certain levels for livelihood and then concentrate on increasing material wealth. We are both consumers and producers in the same breath. Along these lines, the country need not rely exclusively on its natural resource base to increase wealth. The country has plenty of "extra hands" with which to increase productivity and income in line with environmental protection.

Environmental protection connotes both the protection of existing resources and ecosystems and the regeneration of damaged resources. Both protection and improvement require manpower. Under present social, economic, and cultural conditions, China's labor resources may be applied best to large-scale activities in environmental protection and improvements. Projects may include developing resource industries, afforestation, water conservation, and sanitation. Using the vast population for environmental protection efforts will characterize China's approach to the country's many complex problems.

Characteristics Unique to China

The following six characteristics make China's population and environment problems unique: (1) regional diversity, (2) inefficient resource utilization, (3) an urgency to act, (4) sustained increase in productivity, (5) early stages of development, and (6) abundant human resources.

Regional Diversity

Landscape, climate, vegetation, and soil composition are only a few of the parameters that suggest how varied China's different regions really are. The diversities have contributed to uneven stages of development throughout the country. Variable environmental and economic conditions in turn affect the country's population distribution. On alluvial areas of the Yangtze and Pearl River valleys, for example, extensive agriculture, rivers, and ponds provide a wide range of opportunities for livelihood that attract inhabitants. In the Northwest, a relatively sparse population occupies more arid territory. Inhabitants have been attracted wherever opportunities call.

In spite of the uneven distribution, serious environmental problems exist everywhere. Poor industrial and domestic waste management afflicts dense urban centers with intense economic activity. Damage to unique ecosystems is the primary concern in rural areas where natural resource exploitation is rampant. Water shortages and contamination of water sources are commonplace in the North, whereas acid rain afflicts areas in the South. Lastly, urban areas typically suffer from ambient noise pollution, whereas rural areas suffer more from contaminated water and use of pesticides, herbicides, and fertilizers. The range and diversity of regional problems in China is matched by few countries in the world.

Inefficient Resource Utilization

Use of China's resources is inefficient. Any population that exceeds carrying capacity is in a fragile state. Rapid increases in population or large-scale damage to natural resources can spur natural disasters. China's territory covers 9.6 million km^2 and ranks third largest in the world. Population density amounts to 120 persons per km^2. Mountainous areas, plateaus, and deserts form a large proportion of China's territory. These areas have fragile ecosystems and low levels of biological productivity. Although mineral deposits exist, they cannot be rapidly developed into production. Financial shortages and poor transportation infrastructure worsen the low utilization rates.

An Urgency to Act

Immediate action must be taken to steer the country away from even worse problems. Since 1949, China's population has doubled. The increase has slowed the economy and exacerbated the already precarious relationship between people and land. At present, China's childbirth rates are at a new high. By the year 2000, the population will reach 1.3 billion people and continue to grow. This is not the most opportune time for China to cut back on the use of coal as a primary energy source. Air pollution from coal burning will therefore continue, as will the shortage in funds.

Urbanization in areas that lack basic facilities and infrastructure will continue. Intensive agriculture will continue to damage cultivated areas. Margins between the supply and demand of forest products will widen as forested areas and wildlife habitats shrink. Water shortages for industrial and domestic use will also continue. As is clear, China's environmental affairs will unquestionably worsen before they get better. The prospects for the immediate future look grim. The need for appropriate, effective solutions is, hence, urgent.

Sustained Increase in Productivity

Increasing protracted productivity is needed to satisfy the demands of the growing population and help pay for environmental protection measures. In the next two to three generations, China's population is set to balloon. With a relatively young population structure, the majority of women are of childbirth age. It follows that even if strict family-planning policies are implemented, there will be little significant change in the near future. If the implementation of family-planning measures is neglected, however, massive population growth will last still longer into the future.

Decreasing rates of mortality are attributed to several factors. In socialist China, medical services have steadily improved. Epidemics are no longer a pressing concern. Moreover, there is less risk of large-scale war because global politics now emphasize dialogue over confrontation. Even if large-scale war were to erupt, there would be little chance of China's participation due to the nation's "five principles of peaceful coexistence." In sum, there is scant chance for unnatural population declines in China for the foreseeable future. The country had better take measures to prepare for additional growth in numbers.

Early Stages of Development

At its preliminary stage of industrialization, China's developing economy has channeled a large segment of agricultural labor to nonagricultural sectors of the economy. Township enterprises, for example, have risen to fill the gap between agriculture and urban industry. Rapid population growth, however, has precluded any decrease in the aggregate number of unemployed laborers.

Abundant Human Resources

China's labor resources might be converted into an effective force to protect the environment. Industrialization and modernization in other countries suggest that an expanding manpower reserve is an important tool in the socioeconomic development of a country. Why not apply these resources to environmental protection efforts? China's rich manpower resource is its trump card in its goal to protect the environment. Diversity, urgency, and complexity best characterize China's pressing population and environment issues.

13

Controlling Population to Protect the Environment

By the 1960s and 1970s, population control and environmental protection became part of the global agenda. The United Nations convened a series of important conferences: in 1972 Stockholm hosted the Human Environment Conference, and in 1974 Bucharest hosted the World Population Conference.

At the time, China's population exceeded 800 million. Government authorities had already begun to issue instructions that population growth be kept under control. The concern, in addition to widespread pollution, prompted China's first National Environmental Protection Conference in 1973. The convention reemphasized the necessity to act promptly before matters grew worse. In 1974, the National Environmental Protection Agency was established under the State Council. Family planning became an actual pillar of state policies in 1981 with the establishment of the State Family Planning Commission. Environmental protection also became a pillar in the mid-1980s when the State Council established its Environmental Protection Commission.

As is evident, China was quick to act (at least administratively) when population and environmental problems began to reveal themselves. The two issues entered China's political stage simultaneously by no simple coincidence. Legislators were sharply aware that population pressures throughout the 1970s had exacerbated the country's environmental affairs. To reduce population growth and improve environmental affairs, various policy measures have been actively pursued.

A Review of China's Population Policies

Since 1949, China has been free of incessant warfare. A secure environment benefited national reconstruction and set the stage for tremendous population growth. The population has doubled in forty-odd years. High death and low natural growth rates have changed to low death and high natural growth rates. In more recent years, the high birth and natural growth rates have considerably fallen off.

China's family-planning policies have played a tremendous role in this more recent trend. Demographic experts attribute 80% of the success in birthrate reductions to successful government programs. Declines in urban birthrates, in particular, have been remarkable.

China carried out its first population census during the First Five-Year Plan in 1953. It estimated the population at more than 600 million people, including inhabitants of Taiwan, overseas Chinese, and students abroad. Rapid growth and ensuing impacts on economic expansion aroused the attention of mindful planners. Ma Yinchu and other well-known economists advocated birth control and family planning as early as 1957. In *New Demography*, Ma offered the following views on China's population problems.

- Rapid growth and limited government funds have become unfortunate characteristics of China's socialist construction.
- Limited food and industrial raw material supplies are most easily alleviated by population control.
- The country must address both population control and population improvements (e.g., public awareness, education, sanitation, and so on).
- Later marriages and other family-planning practices must be encouraged through economic incentives. Taxes should be levied in accordance with the number of children in a family.
- Population growth should be tailored to the government's development plans.

Ma warned that if China permitted its population to grow unchecked, the problem would certainly hamper—not help—its path of development.

Ma's and others' outcries were unfortunately gagged by subsequent movements in China's modern history. In the late 1950s, Ma's views suffered attack from all corners of the country. One popular view held that continuous population growth conformed to laws of socialism, and that no problems existed. According to this view, the Great Leap Forward was not limited by overpopulation, but by insufficient manpower. A clear dividing line was drawn between those who advocated birth control, called "Malthusians," and those who advocated continued population growth, dubbed "Marxists."

China experienced a Great Decline after its attempted Great Leap Forward, validating Ma's and others' hypotheses. In December 1962, the Central Committee of the Chinese Communist Party (CCP) and the State Council issued the following directives on family planning: (1) implement birth control in urban and densely populated rural areas, (2) determine the appropriate growth rate, and

(3) bring population growth under control accordingly. Propaganda experienced positive results in urban areas but had little impact in the countryside.

A second census in 1964 disclosed a population approaching 700 million people (excluding Taiwan, Hong Kong, Macao, and overseas Chinese). By 1969, the population exceeded 800 million people. China responded by implementing the most innovative family-planning policy in the world.

As directed by Mao Zedong, the Central Committee of the CCP declared that family-planning initiatives were of critical importance. The tables began to turn in favor of the Malthusians. When the State Council established its Family Planning Commission in 1973, population growth indices were introduced for socioeconomic plans. The following guidelines were also established:

- Coordinate population growth and productivity of consumer and capital goods.
- Plan population growth in line with socialism.
- Steer unwieldy population trends to suit development plans.

Although still not fully realized, these guidelines marked a new awareness of the importance of family planning.

Family planning was initially advocated with slogans such as: "Later, sparser, and fewer." The campaigns suggested that later marriages, lengthened intervals between childbirth, and fewer children per family would help reduce population growth. More strict family-planning measures were established in the Fourth Five-Year Plan. Quotas were in fact set for each family, as suggested in the slogan, "One child is not too few, two is just enough, and three is too many." Increasingly strict standards are reflected in more recent slogans, such as: "One is the optimum number and two is the utmost limit."

Birthrates subsequently fell from 6 per family in the 1950s and 1960s to 2.4 in the 1970s. Natural growth rates also fell from 2.583% to 1.187%. Similar achievements in such short time spans are unparalleled among other developing countries.

Both the government and the public became increasingly aware of acute population problems throughout the 1970s. People began to consider that even if each family limited births to 2 children, rapid growth would last another 50 years, at which time pressures on the environment would be unimaginable. In light of this, population policies were further revised in 1979. Calls for a "one child policy" became more frequent. In 1980, the Central Committee of the CCP issued "An Open Letter to All Communist Party members and Communist Youth members on the Control of Population Growth of Our Country." Strict quotas and a one-child policy emerged as the consensus. Conditions were specified for urban versus rural areas, and for minority versus Han Chinese communities. The State Council's Family Planning Commission established the following guideline: "Family planning is a fundamental policy of our country; it should be implemented with persistence." It further mandated that China's population be kept within 1.2 billion by the end of the century.

In 1984, strict family-planning measures were eased to overcome low productivity in rural areas and insufficient care for the elderly. Two-children families were permitted following ratification by local authorities.

Current Population Policies

In the past 20 years, China's population policies have impacted migration, marriage, family, and labor trends. Since the beginning of the 1980s, the government has repeatedly declared family planning a fundamental state policy. Propaganda and education have been central to government efforts because of entrenched cultural views on childbirth. The teeth of the policy is in tax legislation that rewards or penalizes families according to their decisions on childbirth.

Cultural traditions make China unique. In certain circumstances, they also make China overpopulated and poor. Slogans such as "The more sons, the more bliss" and "Five sons are not too many for a man" epitomize aspects of traditional culture that are detrimental to China's progress as a nation. Without awareness and voluntary public participation, family-planning objectives may never be achieved. It is therefore important to express urgency and educate people about family-planning alternatives.

Methods of propaganda and public education have continuously improved. Emphasis has shifted to advice and services, specifically in regard to birth control and contraception. The efforts have become more effective as administrators have learned to tailor their dialogue to specific audiences. In other words, urban households learn about different aspects of China's population problem than do rural households, and northern households hear different advice than southern households. The unmarried, the married, the more affluent, and the poor each define separate sets of problems. The aim of government family planning is to address these groups' individual needs.

Basic principles, control targets, and enforcement are each spelled out in China's constitution and other government plans. The current constitution stipulates that family-planning initiatives are the responsibility of the state and that practicing family planning is the obligation of its citizens. In socioeconomic development plans, control quotas depend on population size and growth rates.

To implement family-planning policies and achieve planned targets, a responsibility system has been established that incorporates different levels of government, grassroots organizations, and married couples. Childbirth quotas, lengthened time intervals between births, and postponement of marriage are each advocated in responsibility contracts between individual couples and their local danwei (neighborhood political organ).

Administrative organizations have been in effect since the early 1970s and grassroots organizations mainly since 1979. At the various levels, and especially at the grassroots, full-time cadres guide family-planning initiatives. Mass organizations—namely, the Women's Federation, trade unions, and the Communist Youth League—play active roles. Government social security services additionally provide health care, insurance, and provisions for the elderly. It is hoped that a sound social security system will alleviate pressures to have children for the sake of elderly care.

Financial incentives have played a large role in reducing childbirth to manageable levels. Families that conform to government directives receive financial support. Families with a single child, for example, are granted a regular allowance by the state and receive preferential treatment with regard to kindergarten and primary school admissions. Families that do not adhere to directives incur penalties as well as social restrictions. The penalty might be a reduced wage or bonus.

A wide range of birth-control measures, specifically contraception, are advocated by the state. Compliance is, of course, voluntary. Measures that have proved popular and effective include use of intrauterine devices and sterilization; these are commonly provided free of charge. The measures are convenient and they safeguard people's rights. The majority of couples of child-bearing age have adopted long-term contraceptive devices that have had a great impact in reducing overall growth rates.

A 1981 Marriage Law prohibits the following: (1) marriage between close relatives, (2) marriage by individuals infected with chronic disease, and (3) early marriage. Medicine has complemented implementation of the law with new information on genetics, premarital check-ups, and childbirth care.

Achievements and Failures

China's population policies are for the fundamental, long-term interests of its citizens. Unremitting efforts have been devoted to establishing and implementing these policies. Achievements during the past 20 years are becoming universally acknowledged.

Since 1949, death rates have fallen considerably. They currently range from 6–9%. Reductions in birth rates are, hence, the key to controlling population growth. Family planning has so far enjoyed tremendous success (see Figure 13.1). Birthrates have fallen from 5.81 in 1970 to 2.24 in 1980; they leveled at 2.5 children per family in 1990. A primary reason for the slight increase in birthrates in the late 1980s is that couples born during the baby boom of the 1960s reached child-bearing ages in the late 1980s. As one-child families become more prevalent, the trends should continue to improve. In the past few years, natural population growth rates have remained at 1.4%—lower than similar figures in other developing countries and lower than the world average.

Conventional ideas on childbirth, specifically the importance of sons, have changed as voluntary compliance with family planning becomes more important

Figure 13.1 (a) Aggregate Birthrate, 1973–1989; (b) Increasing Compliance with One-Child Families

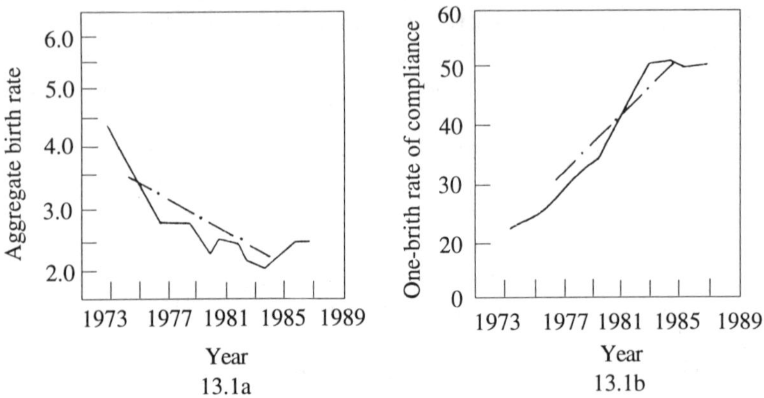

and ingrained. Surveys indicate that most peasants hope to have 2 children. Inhabitants of remote border regions and minority groups in China also generally intend to have more children than those near urban areas. Sterilization appears to be increasingly common.

Population-control measures stand to gain tremendous returns for the country. A decrease of 200 million newborns saves an estimated RMB 2 trillion yuan of government spending for current social welfare programs. Moreover, population control will alleviate demands on China's limited resources. Added grains, raw materials, energy, transport, municipal services, education, housing, and employment opportunity will alleviate increasing pressures on the environment and contribute to national stability and productivity. Per capita national income should also increase correspondingly.

China still has an untold number of obstacles. A large population base from the 1960s caused new peaks in childbirth rates. Even if current population policies are implemented and enforced, the population is due to exceed 1.25 billion and possibly 1.3 billion by the year 2000. Pressures on the environment and limitations on China's economic expansion will continue. Housing, educational opportunities, employment, municipal facilities, and resource availability will each continue to be in short supply. Improvements in China's environmental affairs may stagnate.

At the country's present stage of development, the large proportion of rural inhabitants will not rapidly decline. Plagued with financial difficulties and heavily influenced by traditional ideas on childbirth, rural inhabitants generally have a more difficult time with government directives than do urban inhabitants. Increasingly severe restrictions on childbirth have aroused some public resistance. The fact that total birthrates have remained at 2.4 children per family in recent years suggests that the tremendous progress since the mid-1970s has stagnated some.

Inadequate funding and poor public awareness inhibit rural population problems from being solved anytime soon. Surveys indicate that less-educated couples generally have families above government quotas. Investments in education should, hence, be a priority in the countryside. The rapidity of change has also caused problems. Care for the elderly will be expensive considering China's early stage of development. By the year 2025, the total number of people older than 60 years will increase to 300 million, or 20.3% of the population. An estimated 10% will be regarded as senile.

Environmental Policies

Since the Industrial Revolution, humankind's impact on the environment has grown considerably. The modern world's free market economies have yet to fully cooperate to repair damages from their industrial growth. Economic development in the West created serious environmental pollution that went unchecked until public awareness and environmental protection movements rose to meet the challenge. Since then, strict policies and regulation have had considerable success in the cleanup. International cooperative efforts have made headway, in spite of domestic political debates and the watering down of promises made abroad. Since the 1970s, environmental damage in Eastern Europe has indicated that command economies also fail to curb damage to the environment.

A sound understanding of environmental conditions in countries mandates knowledge of past policies in those countries. A fine line deciphers the impact of environmental policies from population policies. Changes in population and the environment can, furthermore, be spontaneous and beyond state control.

After 1949, environmental damage in China from protracted wars and social chaos ended. Environmental affairs began to take a turn for the better. Some problems still stem from impoverishment and underdevelopment and will solve themselves as the economy develops. The country is now primarily concerned with resolving environmental damage caused by poor policies. How can China pursue its development goals without causing massive degradation to its natural resource base?

In the past 40 years, China's environment has undergone a multitude of changes. As with population control, regulations have tended to be post facto. They react to damage already inflicted instead of preventing future damage. Prior to the 1970s, few policies concerning water and soil conservation and forest preserves even existed. The early 1970s were watershed years in this regard.

Environmental protection was absent from China's First Five-Year Plan. Certain progress was achieved, however, in regard to the construction of municipal facilities, general improvements in sanitation in urban areas, and the protection of rivers. Documents such as "Provisional Hygiene Standards for the Designing of Industrial Enterprises" and "The Temporary Program of Water and Soil Conservation" began to circulate in government administrative offices. In 1958, guidelines on economic development emphasized growth in domestic steel production above all else. Nearly all existing regulations were breached in achieving growth in steel production. Intensive mining and deforestation to fuel steel furnaces cause serious environmental damage. China's increasingly responsible natural resource managers took a back seat to rapid growth targets.

At the beginning of the 1960s, government reforms called for "readjustment, consolidation, restoration, and high standards." Many industrial projects were shelved as China concentrated on the industries at hand. The government made attempts to redistribute industry more evenly across the country and strengthen resource management in general. The government passed and implemented Regulations Concerning Forest Management in 1963 as well.

During the 1966 Cultural Revolution, the few existing rules and regulations concerning environmental protection were labeled "capitalistic, inhibiting, and revisionist" and subsequently abrogated. From this period until 1972, China's environmental movement quieted, as degradation and pollution spread rampantly. Serious environmental dilemmas surfaced one after the other in the early 1970s such that the government had no option but to take immediate action. At this critical juncture, the United Nations convened its Human Environment Congress in Stockholm in May 1972. The congress marked tremendous progress in international efforts to protect the environment. It also marked a turning point in China's domestic environmental movement.

In 1973, the State Council convened the first National Environmental Protection Conference. The conference held that sound environmental protection could wait no longer. The congress subsequently passed Regulations Concerning Protection and Improvement of the Environment, which stipulated rational, comprehensive use of resources and public participation, among other things. After approval, the congress transferred regulatory measures, implementation, and enforcement responsibilities to lower levels of government. It further emphasized that programs be promptly implemented for urban areas, harbors, polluted rivers, industry, and mining operations. Funding for materials and equipment was to take priority in appropriate central and local budgets.

In May 1974, the State Council established a Leading Group on Environmental Protection. The body is primarily comprised of representatives from concerned ministries of government. Among other functions, the group has helped to establish research and monitoring stations in various localities. Its efforts have helped to launch assessments and treatment of pollution. Although pollution has been mitigated in some areas, it has grown worse in the aggregate.

Since 1977, the government has considered environmental protection an important component of socialist modernization. The Constitution of the People's Republic of China in fact stipulated in February 1978 that "the state protect the environment and prevent pollution and other public hazards." In 1979, China's

Environmental Protection Law set out additional rules and regulations. The National Environmental Protection Agency formulates, enforces, and supervises additional guidelines for environmental protection.

Incomplete understanding and insufficient funding are two primary obstacles to implementing environmental protection. Policymakers are forced to decide on the highest and best use of money when it becomes available. Making priorities as such is difficult work in light of how interconnected environmental affairs really are. Current funding for environmental protection is limited. China has, therefore, explored new avenues for environmental protection that better suit domestic conditions.

Three Guiding Principles for Protection

Prevention before treatment is the first guiding principle, especially for countries with massive development potential such as China. All measures need to be taken in step with economic growth. Rooting out primary pollutant sources to reduce expensive clean-up efforts is equally important. Prevention includes the following: (1) establishing environmental protection in accordance with socioeconomic development; (2) conducting environmental impact studies prior to new development projects; and (3) encouraging domestic suppliers of environmental technology for industry and agriculture.

Since the 1980s, environmental impact statements have already derailed or reoriented projects that would have been detrimental to their surroundings. From 1985 to 1988, regulations on large and medium-scale projects have also succeeded in increasing the country's water treatment capacity by 3.85 million tons per day, waste gas treatment capacity by 113.86 million m^3 per hour, and solid waste treatment capacity by 13.42 million tons per year.

A second guiding principle for environmental protection is to increase the liability and responsibility of polluters. Public agencies and private organizations will increasingly be required to internalize formerly external costs associated with pollution control. This should extend government funds for other environmental protection measures. In 1988, RMB 1.21 billion yuan was allotted to purchasing new pollution-control technology. Pollution-control projects need to follow a realistic time schedule. Since 1987, more than 120,000 projects have been arranged for pollution-control improvements. A number of long-standing pollution problems have been solved. Enterprises whose waste emissions exceed state standards are fined, and payments are invested into a pollution treatment fund. By the end of 1989, the fund had collected RMB 7.65 billion yuan.

A third guiding policy is to strengthen administrative networks. A stronger administrative apparatus will better monitor small-scale polluters and increase compliance with regulations and the payment of fines. China's Environmental Protection Law underwent revisions in 1989. The new legal system is more comprehensive; only a sound administrative network will make it more effective. Government propaganda, public education, and scientific research will also contribute in this regard. The government currently employs approximately 200,000 people nationwide to help with environmental protection measures. Projects in urban areas, not surprisingly, tend to be most cost-effective.

Five management "systems" have recently been put into effect. They include target responsibility, quantitative monitoring, pollution permits, centralized pollution control, and a prioritized timetable. Target responsibility sets goals for government responsibilities at various levels. The goals coordinate environmental protection standards with local development plans, and specify the role of government. The system makes the heads of local governments responsible for projects in their domain. Quantitative monitoring of pollution treatment facilities is a second system to improve environmental protection. Surveys and appraisals have become increasingly widespread in most large cities in the 1990s. Issuing pollution permits has become a third effective system for environmental management in 50 cities including Shanghai and Xuzhou. A certain number of permits are allotted to a given region and then traded between industries as a commodity. The system brings pollution under control by limiting aggregate pollutants in a region and by encouraging pollution-control innovation among competing industries. Centralized pollution control is a fourth management system to more evenly distribute pollution treatment efforts. The system should bring in additional profits to the environmental protection fund by increasing compliance in more remote areas. Lastly, a system that incorporates a time limit will make legislation generally more effective.

Achievements and Failures

Environmental policies have become more effective as they have become increasingly attuned to actual conditions in China's various regions. Improved population control and public awareness as well as more efficient use of natural resources will further help minimize environmental degradation.

In recent years, industrial pollutant emissions have fallen by wide margins as national standards have become more strict. Most boilers, furnaces, and kilns with high energy consumption, low heat efficiency, and high pollution emissions have been overhauled. Atmospheric pollution from coal burning has also become less severe. Recycled wastewater, waste gas, and solid waste has increased profits and mitigated pollution (See Figures 13.2 and 13.3). Urban living standards have improved with increased water supply capacity and water treatment facilities. Use of natural gas and central heating have become more widespread as well.

The government has overseen 2,000 agricultural experiments in organic farming and the impacts of chemicals on soil fertility and produce. Treatment of saline-alkali soils has helped farmers for years. Once-sterile and unproductive areas on the Huang-Huai-Hai Plain have been transformed into highly productive areas for grain. Shelter forests in North China and the Yangtze River Valley

Figure 13.2 Recycled Industrial Wastewater, 1982–1989

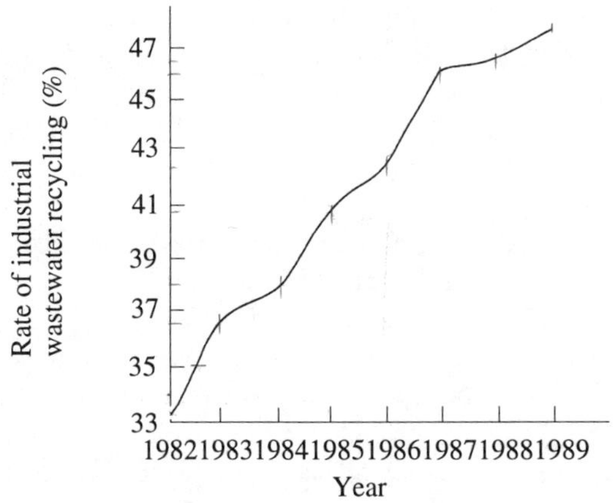

Figure 13.3 Recycled Industrial Solid Waste, 1981–1989

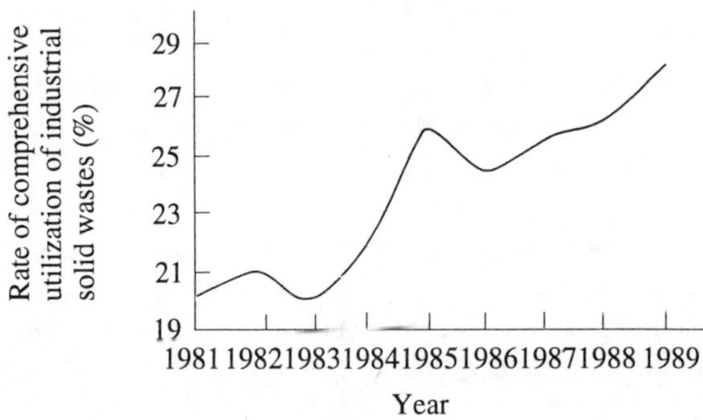

have helped afforestation projects. The Taihangshan Mountain Range has also experienced new, apparently successful forestry techniques. Large-scale water and soil conservation projects have been in operation for several years. Eroded areas under treatment now amount to 520,000 km^2. In North China, prevention and regeneration of desertified areas has increased after years of research.

Preserves

Preserves have been set aside since the 1950s. By 1990, more than 6,000 preserves had been established throughout the country, covering 40 million ha, or 4% of the territory. As hunting, fishing, and timber harvests have subsequently become more limited, zoos and botanical gardens have become increasingly popular. Some areas of the country domesticate species on the verge of extinction; others import and cultivate once-indigenous plants.

Continued efforts like these may be postponed until social stability and development are more secure. Government funding needs to alleviate social pressures from unemployment and housing shortages before addressing potentially more important, long-term concerns such as species and ecosystem extinction. In the meantime, intensive agricultural practices will stress the country's ecosystems. Deforestation and reclamation of grasslands will continue.

As these ecosystems disintegrate, so do the critical roles they play for human livelihood. Natural filtration of water and wildlife diversity are two of the more prominent attributes of forested areas that will suffer.

Weak implementation of protection measures has led to a prevailing opinion that China's environmental policies are ineffective. At present, guiding principles have yet to be fully integrated into overall development strategies. Some of the country's economic strategies still encourage insufficient use of resources. Strategies, standards, and administrative apparatuses leave much to be desired.

Coordinating Population and Environment Policies

Forming Policies

Population control and environmental degradation are closely related. All policies, therefore, that impact one issue must incorporate the other. This has not been the case for the past 20 years. Population and environment policies have typically taken shape independent of one another. If China fails to bring its population under control, the country will not have an adequate resource base for economic development. Controlling pollution and environmental degradation will place the country in better stead to support its population and achieve its ambitious economic objectives.

Carrying Capacity of China's Resources

China's current resource base bears a heavy population load at low levels of economic and technological development. Its agrarian-dominated economy relies on traditional and intensive farming methods. The burgeoning population has exacerbated flooding, soil erosion, desertification, and pollution, among other problems. Population levels far exceed sustainable levels. To alleviate population pressures, growth should be kept at carrying capacity as determined by expert scientists.

As early as 1956, Professor Sun Benwen proposed the appropriate level according to the country's arable land mass at 800 million people. Since the 1970s, many scientists have arrived at similar figures relative to China's food supply. According to U.S. and British food consumption, China's maximum concentration should be 700 to 750 million people. Other scientists have done research on freshwater resources and suggest that the maximum population that China's resources may support is 630 to 650 million people. The range of estimates in fact reach as high as 1 to 1.15 billion people based on the country's development potential.

At present, the population has exceeded 1.1 billion, and is therefore already as great as the highest estimate of China's carrying capacity. Under strict control, the population may peak at 1.5 billion in the twenty-first century. Under slackened control, it may exceed 1.3 billion by the end of this century and peak well beyond 1.5 million later in the next century. Under this more severe scenario, the country's natural resources will be overburdened and the economy will collapse. After a thorough investigation of the country's actual carrying capacity and resource stocks, China should enact measures that reduce and stabilize the population accordingly.

Population Distribution and Migration

Population distribution and migration are functions of the land's carrying capacity. Since the Industrial Revolution, the population has become less dependent on agricultural resources, and therefore less sensitive to locations' carrying capacity. Water availability, soil fertility, climate, and other factors, however, still influence population densities to a large degree.

A rough survey of the resource potential in various regions of China indicates that population distributions in the East and West are in equilibrium. The East is characteristically moist with abundant rainfall in the South and warm or temperate climates. Its produce, timber, and livestock production account for 90% of the country's totals. Its population comprises 95% of the total. Biomass production averages 7 to 20 tons per ha, and each square kilometer bears an average 225 persons. The West is more arid with scant rainfall and temperature extremes. Total available water resources amount to a small 4.6% of the national aggregate; farmland comprises less than 10%. Biomass production averages 1 to 4 tons per km, and each square kilometer bears an average 12 persons.[1] The 1978 UN Conference on Desertification in Nairobi set the carrying capacity of arid regions at 7 persons per km^2, and that of semiarid regions at 20 persons per km^2. Scientific indices as such will help China to formulate more effective environmental policies.

Populations in the Northwest, however, have exceeded scientifically determined limits since 1982. Densities in the area have been as high as 22 persons per km^2. Aside from people trained in science, technology, or other skilled professions, people should not be encouraged to immigrate as this area further develops.

In general, reversing current migratory trends would best suit the environment. Sparsely populated and resource-poor areas are typically in fragile ecosystems. Migration should be oriented toward areas with rich resource potential. As in Africa's Sahara Desert, China's Northwest has a large number of environmental refugees. Migration to more resource-abundant areas will help migrants accumulate wealth and slow environmental degradation in these marginal, fragile areas.

Environmental migration as such has already led to tremendous success. In Gansu Province, an estimated 54,000 people have moved from impoverished Dingxi to Hexi, west of the Yellow River, where natural conditions are more favorable. In Ningxia Hui Autonomous Region, an estimated 78,000 people have moved from impoverished Xihaigu to the more prosperous Band Area of the Yellow River. The migrations have improved living standards and slowed environmental ruin.

Migrations from mountainous areas to the plains and from remote areas to urban centers are expected in large degree and are of benefit to the environment if controlled. In parts of Zhejiang Province, for example, peasants move from the mountains to the plains to practice agriculture, and from the plains to the cities to engage in commercial and industrial activities. These migration patterns have relieved population pressures and improved productivity for rural and urban inhabitants. China should therefore encourage development of township- and village-run industries so that concentrated facilities and infrastructure contribute to pollution control. A primary objective should be to develop a network of urban areas that absorb surplus labor. The task will incorporate population, resource and economic, and environmental concerns.

China's Floating Population

China's rural labor force exceeds 400 million, or 70% of the country's entire labor force. More than 100 million surplus laborers live in the countryside. The government's role should be to help channel excess labor to labor-intensive agriculture and industry, mining operations, and innovative land-use alternatives in nonarable areas. Stable social development and sound natural resource management will continue only if rural surplus labor can effectively be employed in development initiatives. Nonagricultural initiatives currently absorb a small percentage of idle labor resources. Freshly employed labor in rural areas totaled 94.29 million people from 1978 to 1988. During this same period, an estimated 54.61 million people followed nonagricultural pursuits. Employment opportunities simply do not match the growing labor supply throughout China.

Job creation is not without costs to the environment. Mining for mineral resources, in particular, has had serious consequences. Moreover, inadequate funding and technology and a largely unskilled labor force have restrained experiments in organic farming and other environmentally friendly methods of farming. Pilot projects in such fields will take time. Organic farming depends largely on local hydrology, soils, and climate. It is uncertain that productivity will match the variety and quantity of demand in urban and rural areas. Initiatives in organic farming, therefore, should be integrated with industrialization and urbanization.

Using surplus labor in environmental initiatives has been effective in certain areas of China. River valleys of the Loess Plateau, desert areas in the North, and areas with alkaline soils along the coastal plain are a few areas where labor-intensive efforts are in demand. The initiatives require tremendous financial and technological inputs by the state as well. As the benefits of environmental initiatives are long-term and subtle, it will be difficult to maintain interest in these financially burdensome projects. China's administrative apparatuses will feel the strain as well.

It is thus imperative that China redouble its efforts to relieve population pressures and preserve its resources. China should look to environmental protection measures, as well as other development initiatives, to employ its floating population.

Immigration to Urban Areas

Urbanization in developed countries is a gradual migratory process. In developing countries, urbanization has been more sudden. Rapid population growth and industrialization stress urban infrastructure and lead to air pollution, water shortages, and poor solid waste management. Squatter settlements and slums ensue. At current rates of development, no urban development plan could keep pace with the population influx from rural areas. Furthermore, outright control of the rural population by means of legislation is unjust and impractical. An effective measure is to channel migrants toward medium-sized and small cities.

Between 1949 and 1982, the percentage of urban inhabitants in the population was small in spite of rapid population growth. Growth in small towns was generally stagnant, whereas growth in medium-sized and large cities was generally greater. In the early and mid-1980s, rural inhabitants flocked to both small and large urban areas. The migration alleviated pressures in large areas by more evenly spreading the distribution. Guiding this process with government incentives will ease urban degradation and infrastructure failures.

Funding, Technology, and Efficient Use of Resources

Shortages in funding and accessible resources will persist. China must therefore steer away from environmental projects that focus exclusively on large investment projects. The fundamental approach should be to improve the efficiency of resource use by reducing consumption and adopting low-cost pollution-control technologies. Moreover, China should use its competitive advantage in labor supply. Environmental projects in rural areas, urban afforestation, and recycling are a few labor-intensive programs that might benefit.

Note

1. *Subsistence and Development,* p. 29.

14

From Survival to Sustained Development

Since the 1980s, China's development strategy has coordinated population control, environmental protection, and economic development. Some efforts have succeeded and others have failed; all have been painstaking. Certain problems continue to threaten China's survival and growth. Refining development strategies to match the country's unique conditions is the key to China's success.

General Strategies

China faces the following major constraints: population growth, pollution, inaccessible or dwindling resource stocks, and inadequate government financing. First, annual net increases in population average more than 15 million people. Even if current population policies are enforced, the population will reach 1.25 billion by the end of the century. Second, preventing environmental pollution will be no simple task. Acid rain, depletion of the ozone layer, transboundary hazardous chemical wastes, and the extinction of plant and animal species all afflict China. Third, per capita availability of resources is bound to further decline as the demand for resources in general increases. Although advances in science and technology will alleviate pressures on China's resource stocks, impacts from growth will increase at faster rates. Fourth, financing is in short supply. China's three-stage strategy of economic development involves first achieving subsistence, then increasing affluence, and finally investing in modernization. Economic strength in the short term will remain low. No single area of the economy will progress without development in other areas. Strategies that call for rapid development in isolated segments of the economy are doomed to failure. Industrial growth, for example, relies on improvements in the country's transportation infrastructure.

The government all too often attributes environmental problems to economic growth. It fails to fully appreciate the potential in investments in science and technology. China must consider all options and all aspects of its development strategy. Strict rural family-planning policies, for example, do not always conform to current economic reforms. Under the contract responsibility system, land is allocated according to household size. Larger families are consequently allocated more land and receive more benefits than smaller families. The policy, hence, encourages local farmers to have more children. A more comprehensive strategy would take into account all aspects of development. Strategies as such are popularly referred to as "strategies of sustainable development."

Broadly defined, sustainable development is a development pattern that satisfies the demands of the current generation without jeopardizing supplies for future generations. Inherent in the definition are two assumptions. First, the needs of the second generation are universal for rich and poor. Second, the environment will be able to support future generations if we are only conscious of current resource use.

Defining levels of sustainable development is an important first step for developed and developing nations alike. International dialogue should be an important vehicle to determine such levels. Major issues in international relations throughout the 1980s included the North-South dialogue, disarmament, development, and the environment. The UN General Assembly, in fact, established three panels to confront these issues. They were headed by then West German Chancellor Willy Brandt, Swedish Prime Minister Olof Palme, and Norwegian Prime Minister Brundtland. The panels each produced documents entitled "Our Common Crisis," "Our Common Security," and "Our Common Future."

Sustainable development has a direct bearing on productivity and general livelihood. It will determine progress and the very survival of the human race. The concept has attracted worldwide attention and is now accepted by most governments and international organizations.

Sustainable development generally presumes the following on development:

- Basic needs for the population include rights to education, employment, and social security in addition to food, clothing, housing, and transportation.
- Growth depends upon restructuring industry, employment, and consumption so that general productivity is made more efficient.

A sustainable development strategy might consist of the following eight elements: (1) population control in terms of quantity as well as "quality" (e.g., literacy, medicine, general education); (2) protected resources with increased utilization rates; (3) pollution control to protect people's health, promote socioeconomic development, and protect the environment; (4) efficient use of resources with reduced per capita consumption and recycling; (5) labor and service industries that ensure social security and equitably distribute income; (6) strong research and development for innovative and practical production technology; (7) an educational system geared toward advances in technology and economic development; and (8) stable international trade.

A strategy as described would change past patterns of development, characterized by uncontrollable population growth, wasteful use of resources, environmental pollution, poor efficiency, income disparity, backward education, and general autarchy. The new strategy promotes sustainable use of natural resources and underlines the importance of a cleaner environment.

Making the Transition

A common task for countries is to move toward sustainable development. The trend will benefit individual nations as well as the world at large. Developing countries, however, can make the transition to sustainable development only having gone through industrialization. Throughout the 1980s, China made tremendous strides in its development. As China's economy further expands, it will have that much more ability to achieve sustainable development.

In the past 40-odd years, China's development strategy has undergone several major shifts at the macro level. The country's development path was once characterized by disregard for the environment and concentrated investments in specific segments of the economy. Despite the benefits of rapid industrialization, China has been plagued by rapid population growth, resource depletion, pollution, stagnating agricultural productivity, poor technology, inefficiency, backward education, and only gradual improvements in living standards.

These shortcomings became increasingly evident in the 1970s, at which point the country's development strategy focused on improving general efficiency. Production subsequently expanded. More recently, there have been calls for sustainable, coordinated growth. The new strategy shifts emphasis from GNP growth to coordinated social, economic, and environmental development. It is without question still in a nascent stage. Contractual responsibility on occasion still leads to short-term profit-taking, pollution, and other negative externalities. Economic reconstruction, however, has led to a marked reduction in energy and steel consumption per unit GNP. Most energy and material consumption indices are either remaining steady or slowly increasing. The transition to sustainable development will without question prove difficult.

Reform and development in the 1980s have shown that changes in strategy can prove difficult and slow. Political battles are, needless to say, time consuming and draining. If China is to have protracted sustainable development, new strategies and reforms need to continue. The country must guard against impatient and rash behavior; otherwise, economic ruin will lead to social instability and political chaos.

Improving Population Policies

Strict population-control measures have achieved tremendous results for China. The potential for future growth is enormous, however, and any optimism should be guarded. In the past 20 years, population targets have been modified several times. Slogans have changed from "One child is not too few, two is just enough, and three is too many" to "One couple, one child."

The issue remains socially sensitive and divided. A large percentage of couples, for instance, desire two or more children in spite of government incentives to the contrary. Moreover, methods of implementation tend to be oversimplified and quotas far from realistic. The context for public policy has undergone tremendous change as well. As the market economy displaces communes, peasants are less subject to administrative controls. It is also increasingly difficult to further reduce the aggregate birthrate, which has leveled at 2.4%. Growth rates are generally easier to control in urban areas. Families in rural areas, however, are willing to pay heavy penalties for second or third births. As policy evasions become more frequent, implementation measures become more costly.

Population policies have become a focus of attention for the general public. Controversy within the academic community abounds. Policymakers should be aware of general trends, seek counsel from experts, and make pragmatic decisions. A tentative understanding of the relationship between environmental protection and population control has led to the recognition that China's foremost concern is population control. Developing human resources and increasing public awareness should also become priorities. These measures would both help to control population growth and turn a weakness into a strength.

Population policies need to be assessed over long time horizons. It should also be made clear that problems are cyclical in nature. China must address not only current but also future problems. The government is beginning to lean on advice by population experts that China's objective should be zero population growth. Policy phases and scopes need to be more clearly defined. For example, a one-child policy should be enforced until aggregate birthrates approach replacement levels.

China's most effective approach will be to emphasize "key links." For example, multiple-birth families currently outnumber single-birth families. China should therefore concentrate on preventing multiple births in rural areas. The country should also modify targets in light of the new situation resulting from rural reform. New regulating mechanisms should include economics, administration, health care, social security, education, and other fields. Fewer aggregate births, moreover, might become a principal indicator for the success of family-planning initiatives. A better-organized, more efficient network that covers the whole country should be established.

Administrative enforcement as well as economic incentives should be used to enforce goals. Administrative enforcement primarily aims to prevent multiple births. In the case of two births or less, other instruments include public information and economic incentive. Policies that target only specific areas are no longer advised.

Three measures will help implement family planning: legislation, education, and health care. Few countries of the world have their population growth controlled by legislation. This approach is a necessity in China. Various population laws concerning child-rearing and demography should be formed to guarantee the rights and interests of the people. Education on child-rearing should also be provided. Additionally, health care facilities and other institutions should be established in rural areas. Lastly, rural areas should establish a social security system to ensure care of the elderly. A social security fund, to which all members of the community contribute, would help to secure the benefits described above. China should integrate improvements in living standards, namely health care, into the overall process of reform and development.

Improving Natural Resources

Two components of China's population policy are to control the size and improve the characteristics of China's population. Limiting population growth will alleviate pressures on the environment. Improving population characteristics (e.g., education and living standards) will change the nature of environmental degradation.

Working people and farmers, in particular, are the focal points for many policies that ultimately aim to reduce the wanton destruction of resources. The government's "Spark Program," in particular, aims to direct appropriate technology to rural areas. This program, in addition to other organic farming programs, also intends to steer surplus labor toward environmental protection efforts. The past 10 years have shown that considerable social, economic, and environmental benefits can be gained by popularizing scientific and technological knowledge among peasants. Where appropriate technology has been used, rational use of resources has helped to raise income and protect resources. Science and technology should be tested and integrated in rural areas as well as urban ones. Vocational training, managerial skills, and knowledge of ecology appropriate to local conditions should help China's progress and stability.

It would furthermore be in China's long-term interest to focus on public awareness and improve population characteristics. The country is currently full of people who are blind to environmental issues. Environmental awareness should be integrated into the national education program so that the public will be informed of the country's environmental status. They will thus perceive impending crises, as it would be impossible to sustain resource consumption trends. The country should strive for moderate growth and limit resource consumption to a minimum. It should become second nature for citizens to protect their environment, instead of damaging it.

Surplus Labor and Conservation

Surplus labor is a widespread, long-term socioeconomic problem in China. The country should strive to turn this liability into an asset. Surplus labor can act as a motivating force for environmental protection. For example, reforestation projects, grasslands development, water conservation projects, cropland fertility improvement, and recycling all require extensive labor resources.

Two primary constraints make the use of surplus labor difficult. The cost of surplus labor is zero and even negative in theory. This is not, however, true in practice. Surplus labor will become increasingly expensive with increases in per capita income. This sort of labor, then, is not necessarily as economically feasible as it was in the 1950s and 1960s. Moreover, it is difficult to find people willing to organize the work. Furthermore, environmental projects are unique. Benefits are long-term and shared by the public. An effective institutional framework and set incentives will overcome these constraints.

Practical methods to overcome the surplus labor problems include organic farming methods and contracts for environmental projects. Organic farming is labor intensive, efficient, and environmentally sound. Its returns are considerable relative to other agricultural activities. There are now more than 2,000 pilot projects of different types and scales in "eco-agriculture." Government at every level should provide leadership in encouraging organic farming methods.

Projects to reforest or otherwise increase productivity on denuded hills, small basins, and desertified and alkaline areas will help agriculture considerably. If subsidies for land improvement programs are pooled, then there will be less demand on financial and technical resources. Water conservation, reforestation, and road-building projects will further help remote, impoverished areas.

Refining Natural Resource Extraction

Utilization Rates

Since the late 1970s China has promulgated more than 10 laws that concern the environment. In 1988, administrative functions of ministries under the State Council were strengthened. Market mechanisms also gradually impacted use of the country's natural resources. Both the institutional and the legal framework for resource management have hence improved.

Industrialization that emphasized quantity over quality led to low prices for resources and high prices for finished products. Although China's approach may be significantly modified, the situation will assuredly not be remedied overnight. Population growth and increases in per capita resource consumption will continue to influence the country's resource base.

International research on the relationship between raw material consumption and GNP growth shows that if per capita GNP is below US$2,000, increases in demand for raw materials are generally higher than GNP growth. If per capita GNP exceeds US$2,000, however, increases in demand for raw materials markedly decline. For the moment, China's aggregate demand for resources is expected to increase quite rapidly because of population growth and per capita increases in demand. Resources, therefore, need to be efficiently converted into tangible national wealth. Investing in practical technology is one measure that will assuredly resolve inefficiencies, reduce demand, and help stabilize economic growth.

Conservation

Conserving resources is of at least equal importance to expanding a country's accessible resource base. A growing resource base without comparable progress in extraction efficiency will both increase pollutants and rapidly deplete residual resource supplies. The problem has yet to receive sufficient attention.

Conservation and efficiency are too often regarded as secondary concerns. It is the responsibility of government departments at all levels to make them a leading priority. Industry can facilitate the transition by focusing on labor and services over material resources in the production process. Phasing out backward technology and high-resource-consuming industries is a first step. Next, efficiency should be a primary consideration in all investment decisions. Government and industry should give priority to conservation projects. Third, economic policies that encourage resource consumption should be phased out while incentives for conservation are established. Current state incentives must be honored and new ones introduced. Improvements in technology and standards for conservation should be established for each major industrial sector. Metallurgy, chemical, and energy sectors are examples of resource-intense industries that should be targeted first.

Special measures should also be instated to protect resources vital to the national economy and people's livelihood. Protected farming zones and water resources are primary examples. The aforementioned policies should eventually become law.

Ownership

Established property rights for natural resources will improve the efficiency of resource use. In developed countries, land titles, deeds, and tax records specify property ownership. Property rights in China, to the contrary, have not been clearly defined. Resource extraction is hence wasteful.

Above all else, natural resources need to be regarded as a national asset that produces lucrative monetary and nonquantifiable returns. As such, the country's resources need to be administered effectively to protect and guarantee the lawful rights of owners and users. Surveys, laws, enforcement, accounting, and financial assessments are each a component of sound administration.

In a planned socialist market economy, various sectors are involved in resource management and use—state-owned, collective, private, and foreign. As the principal owner of China's national resources, the state is entitled to compensation from users for depletion. Acquisitions are also made from the state. Costs are based on value and set to promote rational use. Taxes, commonly levied on large profits, are meant to nationalize some of the returns and equitably distribute benefits in spite of geographic location. The idea is that this will facilitate fair competition. The rights and interests of the state as owner are also primary concerns.

Pricing and the Free Market

In addition to more clearly defining property rights, China needs to revise its overly bureaucratic resource allocation procedure, rationalize pricing, cultivate markets, and control excess demand. The current situation is characterized by "high prices for finished products, low prices for raw materials, and no price for resources." Such a pricing system does not account for factors such as land rents for natural resources. As a result, prices for mineral and forest products are low and price ratios are seriously distorted. Imported fertilizers, sodium carbonate, sulfur, and other products are heavily subsidized. Mining by-products are purchased at artificially low prices. The scenario keeps nominal returns from basic raw material industries at artificially low levels. The whole industry consequently runs at a deficit. Needless to say, the situation has a negative impact on employee morale and hampers technical innovation.

The establishment and operation of a market for resources has been made difficult with the extremely low prices for goods. Low prices and weak market infrastructure have also made it difficult to use the market to influence consumer behavior. The inability of previous administrations to recognize the importance of price in making the use of resources more efficient has contributed to the country's environmental crises. Accurate pricing is a powerful variable in promoting a raw material industry and conserving resources. Economic studies on natural resource valuation should be encouraged. Along these lines, short- and medium-term plans should adjust prices in the raw materials sector. To avoid excessive and immediate price increases, prices on finished goods should be freed incrementally.

Cultivating a market for resource products is the country's next step. Markets will necessarily be diverse and many-tiered. Commodity exchanges for farm, forest, and mining products would also benefit if a futures market evolved. To avoid the chaos associated with these markets, state authorities need to develop laws and regulations that define operating rules.

Developing a Resource Industry

In addition to reducing consumption, China should take measures to protect, rehabilitate, and regenerate natural resources. The country is currently plagued by insufficient investments in preservation and poor resource regeneration. Funding for regeneration efforts such as reforestation are limited in range and efficiency. An effective measure to increase the accessibility of resources and alleviate population pressures is to develop an industry for resource regeneration. Introducing concepts such as depletion and other accounting tools may further help China's state of affairs. Government should take the following measures to build a responsible natural resource industry:

- *Pricing*. Whereas policy, finances, and technology are the primary tools for impacting natural resource management, no single factor is as influential as pricing and the economy. A planned commodity economy transforms resources into commodities and increases operating efficiency.
- *Free markets*. Markets best approximate the value of resources. Despite long cycles and high risk, natural resources are tangible. They should, however, be exchanged in ways unique to ordinary commodities. Benefits of resources are not entirely reflected by myopic markets. They have multiple long-term, public benefits not incorporated in the price.
- *Direct payments*. Values of mineral resource deposits and plantation forests are best established by direct links between suppliers and consumers.
- *Indirect payments*. Payments through industrial development foundations or relevant state authorities are made in two ways. First, payments can be made as reserves. A primary responsibility for the resource industry is to set aside reserves in advance for the state. All production activities should, therefore, be first under the guidance of the state regardless of ownership. Second, payments are also made by the state in fixed sums. These sums are paid by foundations or government branches for projects that benefit the general public (e.g., geological surveys, shelter forests, oceanographic surveys). Alternatively, resource industries may submit projects for approval in line with the State Project Guide and apply for fixed amounts.
- *Resource investments*. Investments should no longer be distributed by sector but through specialized national banks or investment companies. Intermediary bodies as such will be independent of government departments. They would serve the whole industry and be responsible for overseeing projects and evaluating returns.

Natural Resource Accounting

National Economic Accounting (NEA) provides comprehensive information for analyzing economic performance. It fails, however, to account for natural resources so that losses are not reflected. Increases in income fail to account for depletion. As a result, a country that registers increases in GNP or national income may at the same time be exhausting its resources and potential for future growth—clearing forests, eroding arable land, polluting water, destroying wildlife, and over-fishing marine areas.

Natural resources are the foundation for growth in GNP and national income. They also contribute to social stability. If depleted, resources will negatively impact the economy and the country's social stability. Loss of natural resources should, therefore, be depreciated—as with factory buildings, machinery, and other equipment—and then integrated into national accounts. An increasing number of experts have recognized that countries need to follow the dynamic changes in their resource stocks if they intend to follow sustained economic growth.

Calculating resource availability—its value and flow—will help measure aggregate domestic wealth. Over time, the measurements will additionally help define the relationship between resource availability and economic development, thereby aiding resource management decisions in the future. Resource accounting will, moreover, help to define property rights.

The annual decline in natural resource stocking, including extraction and depreciation, should be reflected in monetary terms as capital losses under the liabilities column in national accounts. Annual increases in natural resource stocking—including growth, recently identified reserves, and appreciation—should also be reflected in monetary terms as capital formation under the assets column in national accounts. Accounts as such will reflect the input-output relationship between development and the environment.

Enhancing Environmental Policies

The Individual

Environmental protection in a country with 1.1 billion people is more difficult than protection in countries with comparable conditions and a smaller population. China's approach is therefore unique compared to other countries. Since the country's population is not expected to change dramatically, how should China mold effective environmental policies?

An underlying principle of China's environmental policy is to protect the rights and interests of present and future generations. The government unequivocally aims to ensure that its population live in favorable circumstances. Policies have been designed not only to address problems in areas affected by pollution, but also to improve living standards nationwide. In light of current degradation, priorities need to be given to areas where the largest number of people are impacted.

Living standards are impacted in two ways. Decentralized heating, car exhaust, and soil erosion are each examples of small-scale anthropogenic pollution. Massive effluent discharge, waste gas, and other hazardous materials are examples of large, industrial-scale pollutants. In the case of the first kind of pollution, China's policy has been to mobilize local authorities and reform urban infrastructure. The second kind of pollution is mainly point-source pollution and is more easily controlled. The government needs to strengthen and improve environmental impact statements for major projects. Environmental protection organs have surveyed major pollution sources and have prioritized spending on enforcement according to three distinct degrees of violation. The first group of violators number 3,000 and produce 65% of the total volume of industrial emissions.

Public Awareness

Experiences in other countries show that public participation is a fundamental component of environmental protection. Acting environmentally responsible is an effective deterrent to polluters. It also gives impetus to government efforts to protect the environment.

At present, environmental protection projects are in most instances carried out by the government. When people's environmental interests are violated, they commonly turn to administrative departments for help. They seldom resort to law and use personal means to resolve disputes. Should the country continue to rely on government management or should the government do all it can to help people play a part in environmental protection?

The wide range of environmental concerns and China's huge population make the former alternative unrealistic. The government needs to give more credit to the benefits of environmental awareness. The public will help to safeguard their interests if they simply become aware of current degradation. The most effective way to mobilize the masses is to help create an environmental culture, or a set of values toward the environment. Since the birth of humankind and through the middle of the twentieth century, our "environmental culture" has been characterized by a fear of subordination to nature. Fear was followed by a desire to conquer. Since the 1960s, when environmental crises began to surface, a new environmental culture began. In developed countries, people's income reached high levels and a social attitude toward the environment began to take shape. In light of China's population problem and its current environmental affairs, society cannot simply sit and wait for the emergence of an environmental culture. To the contrary, government and the media need to encourage this culture to take shape.

Environmental protection organs in China have currently emphasized the use of books, journals, films, television, and various other media to popularize scientific knowledge and legislation on the environment. It will be far more difficult to convince individual citizens of their responsibility in the matter. China's traditional utilitarian attitude views the environment in material terms only. A more responsible view of the environment should impact standards of production and patterns of consumption.

If successful, the change in thinking will provide China with a broad foundation for future environmental efforts. The change should prevent degradation, create markets for environmentally sound products, and develop a "green" resource industry.

An Investment in People's Standard of Living

In the final analysis, the fundamental objectives of investments in environmental protection are twofold: to preserve living standards and to further improve the amenities of the environment. Developed countries began to strengthen controls and increase their investments in environmental protection after widespread pollution in the 1950s and 1960s. The investments supplemented strict standards and legal instruments. By the mid-1970s, investments reached a peak 2% of GNP. After enterprises were installed with pollution-control equipment, investment levels dropped back down to 1%. Having resolved many of the pollution problems, developed countries continue to work for better environmental quality. Enterprises are meanwhile pressed to conform to new objectives and increase investments in environmental technology. The percentage of dollars in protection will be between 5% and 6% by 2020 (Figure 14.1).

To ensure that China's investments in environmental protection keep pace with present growth, they must be carefully guided. The government has formulated the following long-term objectives. By the end of this century, pollution will be brought under control and environmental quality in major cities will markedly improve. As a result, coordinated efforts between the environment and socioeconomic development will continue into the future.

Figure 14.1 Developed Countries' Investments in the Environment

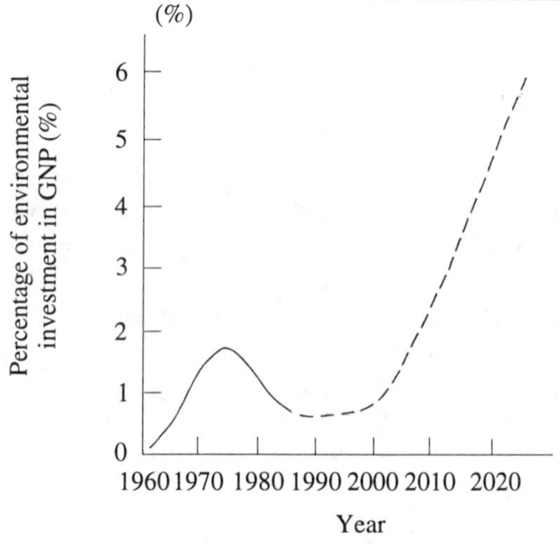

The strategy is two-staged. First, China will aim to preserve living standards. Next, the country will further invest in environmental amenities. To preserve living standards, China's investment in environmental protection should reach 1% of GNP. This percentage will allow unrestrained economic growth with protective measures. Environmental investments have steadily increased from less than RMB 2 billion yuan in the early 1980s to RMB 10.2 billion yuan in 1989. The proportion of environmental investments in GNP has reached 0.7%, still short of the optimal 1%. China should further implement the following measures:

- Incorporate environmental protection indicators into the group of socioeconomic development indicators. Furthermore, incorporate investments in environmental projects (e.g., large- and medium-scale pollution control projects, ecosystem protection projects, and others) into an overall national investment plan.
- Identify and expand sources of financing for government projects. Establish rules and regulations concerning financing to effectively control investments. Taxes on, for example, sulfur emissions might also bring in new sources of funding.
- Match high environmental standards by efforts at implementation and enforcement. Environmental protection, for example, should improve pollution emission standards at national levels and apply more stringent standards to new pollution sources.
- Earmark a proportion of GNP and bank loan quotas for environmental projects, because technology and funding for environmental protection departments are in short supply. Interest rates for such loans should be kept low. Environmental projects should also be exempt from taxes levied on other industries that are heavy users of state infrastructure, such as the Energy-Transport Construction Fund.
- Institute a favorable investment environment and low inflation, which will attract worldwide support.

Science and Technology

Scientific and technological progress is important primarily for two reasons. First, technological breakthroughs can increase the accessibility and supply of natural resources. It increases the utilization rate of resources and reduces material consumption and discharge. Science and technology additionally expand inhabitable areas by enlarging the carrying capacity of the planet. Progress in science and technology will significantly alleviate population pressures on the environment.

Although China formed management guidelines in the 1980s, environmental management at the time was weak. Ten years have since passed, and management has been slow to improve. Excessive use, backward technology, investment shortages, and poor returns have become more acute. Experience in many areas has shown that it is difficult to control degradation and improve environmental quality with weak technology.

China's environmental sciences and technology were first developed in the early 1970s and have since made phenomenal progress. At present, a network of environmental research institutions has been established. The scope of research has hence expanded considerably. Understanding of the country's environment has come a long way. Progress has been achieved in certain technologies, and research in environmental protection, planning, and administration abounds. Overall, however, China is still relatively backward in environmental sciences. Practical pollution-control technology, in particular, is unavailable and in high demand. Without practical technologies, very little of China's research is applied. Science and technology have not played as important a role in environmental protection as they could.

To rectify the situation, environmental protection organs should continue to become more involved in the planning, direction, and supervision of basic and applied research. In the past, their role has stagnated in management. Considering measures in the Eighth and Ninth Five-Year Plans, it is imperative to shift the emphasis from administrative work to more practical concerns of applied research and technology. The country needs to focus on identifying and adopting a number of advanced, efficient technologies. Mandatory use, guided use, and free market dissemination are three effective ways to spread technologies, depending on the circumstance. Technology for industrial pollution control, in particular, should be integrated with the state's industrial policy, technical renovation programs, and state science and technological extension programs. Technology to improve environmental affairs should be integrated with various state agricultural scientific developments. Programs that promote developments in agriculture, such as the Spark Program, are practical technical extension programs.

Bibliography

Cai Wenmei. 1989. "How to Settle the Population Problems in China." *Science and Technology Guide*, no. 4.

Chen Qi, et al. 1988. *China's Natural Resources in 2000*. Shanghai: Shanghai People's Press.

Chen Wei. 1989. "Moving Towards the Coordinated Development of Population, Ecology, and Economy." Paper presented at the First National Symposium on Population and Ecology in Nanjing, 1988.

China Sciences Academy. National Conditions Group. 1989. *Survival and Development*. Beijing: Science Press.

China's Environmental Science Institute. 1989. "A Collection of Essays from the Sino-American Academic Symposium on Environmental Law." Beijing: China Environmental Science Press.

Chinese Science Academy, Commission on China's Physical Geography. 1985. *A Survey of China's Physical Geography*. Beijing: Science Press.

The Chinese Society of Environmental Management, Economics, and Law. *On Environmental Economics*. Nanjing: Jiangsu Science and Technology Science.

College Demographical Research and Training Office of the National Education Commission. 1989. *Exploration and Demography*. Beijing: Beijing University Press.

Fang Rukang. 1989. *Natural Resources in Our Country and Their Rational Use*. Beijing: Science Press.

Feng Jiuzhang. 1989. "Preliminary Studies on the Basic Characteristics of Our Country's Population and Environmental Problems." Paper presented at the First National Symposium on Population and Ecology in Nanjing, 1988.

Foreign Population Series. 1989. Beijing: Research Center for Chinese Population Information.

Fu Lixun, et al. 1989. *Improving the Environment*. Beijing: Academic Books Press.

Gan Zeguang, et al. 1987. *An Introduction to Environmental Economics*. Xi'an: Northwest Industry University Press.

Gao Qun, ed. 1991. *Coordinated Ecological, Environmental, Social, and Economic Development*. Changchun: Jie Lin University Press.

Gu Baochang. 1991. "The Development and Utilization of Birth Data." Beijing: Center for Chinese Population Information.

Gu Songjian. 1989. *An Introduction to the West's Population and Economy*. Beijing: Beijing University Press.

Gui Shiyong, et al. 1988. *On the Socioeconomic Development of Our Country*. Beijing: China Plan Press.

Guo Daping. 1989. "Measures to Control Multiple Birthrates in Rural Areas." *Science and Technology Guide*, no. 4.

Guo Xiaoming, ed. 1990. *Projections on China's Environment in 2000 and Its Countermeasures*. Beijing: Qinghua University Press.

Guo Xiaoming, et al. 1989. "A Quantitative Analysis of China's Economy, Its Population, and the Environment in 2000." *Environmental Science Research*, no. 5.

Han Guogang, ed. 1989. *Save China*. Beijing: Qiu Shi Press.

He Qinglian. 1988. *Population: The Sword Over China*. Chengdu: Sichuan People's Press.

He Zhongkun. 1989. "Population Problems and Countermeasures in Structural Reforms." *Population Front*, no. 2.

He Zhongkun, et al. 1989. "On Strategies to Realize the Coordinated Development of Population and the Environment in China." Paper presented at the First National Symposium on Population and Ecology in Nanjing, 1988.

Hou Dongmin. 1988. "Constraints of Resource Availability on China's Future Development." Paper presented at the First National Symposium on Population and Ecology in Nanjing, 1988.

Hu Angang. 1989. *Population and Development*. Hangzhou: Zhejiang People's Press.

———. 1991. "Burden and Wealth: Exploitation of China's Labor Resources." Paper presented at the Seminar on Resource Potential and Measures for Efficient Use in Beijing, 1990.

Hu Changhai. 1989. *A Geography of China's Forests*. Beijing: Agriculture Press.

Huang Li. 1989. "On the Scientific and Cultural Level of Populations and the Ensuing Benefits." Paper presented at the First National Symposium on Population and Ecology in Nanjing, 1988.

Jiang Fenglan, et al. 1989. "Studies on Environmental Planning and Environmental Protection Targets for 2000." *Environmental Science Research*, no. 6.

Jiang Zhenghua. 1988. "Population Growth and Economic Development in China." *Population and the Economy*, no. 6.

Jin Wenyong, et al. 1989. *Contemporary World Environment*. Beijing: China Environmental Science Press.

Li Chengxun. 1989. *Strategy of Socialist Development*. Nanchang: Jiangxi People's Press.

Li Jinchang. 1988. *Our Resources and Environment*. Beijing: Xinhua Press.

———. 1991a. "Natural Resources Are Not Without Prices." In *Development and Choices: An Analysis of China's Development in 1990*. Beijing: Scientific and Technological Literature Press.

———. 1991b. *On Resource Accounting*. Beijing: Marine Sciences Press.

———. 1991c. "Studies on Measures to Coordinate Economic Development and the Environment." *Management World*, no. 5.

———. 1991d. "Theories on the Value of Natural Resources and Price Fixing." *China's Population, Resources, and Environment*, initial issue, 1991.

———. 1992a. "Rely on Science and Technology, Protect the Environment." *Modernization*, no. 3.

———. 1992b. "Selected Articles on the Environment, Ecology, and Resources." Beijing: China Environmental Science Press.

Li Jinchang, ed. 1990. *An Approach to Resource Accounting*. Beijing: China Environmental Science Press.

Li Jinchang and Zhong Weizhi, eds. 1990. *Theory on Resource Industry*. Beijing: China Environmental Science Press.

Li Qiang. 1989. "China Faces the Crises of An Aging Population." *Science and Technology Guide*, no. 4.
Li Rongshi. 1989. "Analysis of Our Country's Population in 1987." *Population Research*, no. 1.
Li Shuqin. 1989. "The Relationship and Coordinated Development of Population Characteristics and the Environment." Paper presented at the First National Symposium on Population and Ecology in Nanjing, 1988.
Liang Ziming. 1991. *Problems of Population, Labor, and Employment*. Guangzhou: Jinan University Press.
Liu Hong, ed. 1990. *National Conditions in China*. Beijing: Central Party School Press.
Liu Zheng. 1990. *Studies on Problems with China's Population*. Beijing: People's University of China Press.
Lu Zhiqing, 1988. *Strategic Decisions on Population and Development*. Beijing: Xinhua Press.
Luo Qimeng. 1989. "Population, Farmland, and Food." *Population*, no. 3.
Ma Hong, ed. 1988. *China in 2000*. Beijing: China Social Sciences Press.
Ma Xiaoliu. 1989. "Overpopulation and Environmental Degradation." Paper presented at the First National Symposium on Population and Ecology in Nanjing, 1988.
Ma Yintong. 1990. "The Past and Future for China's Population." *Chinese Population Newspaper*, March 2, 1990.
"Make Unremitting Efforts to Control Population Growth." 1989. *Population and the Economy*, no. 3.
Mao Kuangsheng, et al. 1989. *Demographic Principles*. Beijing: China Finance and Economy Press.
National Environmental Protection Agency. 1988–1990. "Annual Statistical Bulletins on the Environment—1988, 1989, 1990." Beijing: China Environmental Science Press.
National Environmental Protection Agency, Chinese Science Academy, Changchun Research Institute of Geography. 1989. *Atlas of Nature Protection in China*. Beijing: Science Press.
National Environmental Protection Agency, Propaganda and Education Department. 1988. *Rivers Won't Flow Forever*. Beijing: China Science Press.
Pan Jiyi. 1988. *Population Ecology*. Shanghai: Fudan University Press.
"Program on Nature Protection in China." 1987. Beijing: China Environmental Science Press.
Qu Geping. 1983. "Environmental Science." In *China Encyclopedia*. Beijing: China Encyclopedia Press.
———. 1984a. *Environmental Problems in China and Policies for Them*. Beijing: China Environmental Science Press.
———. 1984b. "The Impact of Sharp Population Increases on the Environment." Selected articles from the Tenth Anniversary issue of *Environmental Protection*. pp. 104–116.
———. 1985. *On the Synchronized Development of Environmental, Economic, and Social Fields*. Beijing: Marine Sciences Press.
———. 1988. *China's Environmental Protection in 2000*. Beijing: China Environmental Science Press.
———. 1989. *Environmental Management in China*. Beijing: China Environmental Science Press.
———. 1991. "Challenges of China's Population and Environment Problems and the Ways to Meet Them." *China's Population, Resources and Environment*, initial issue, 1991.
Qu Geping, ed. 1988. *Environmental Science Research in China*. Shanghai: Shanghai Science and Technology Press.

Qu Geping and Li Jinchang. 1991. "The Impact of China's Current Population and Coordinated Countermeasures." *China's Environment*, April 4, 1991.
Qu Geping and Li Jinchang, eds. 1988. *China's Environment in 2000*. Beijing: China Social Sciences Press.
Qu Geping and San Yichu. 1988. *The Development of World Environmental Problems*. Beijing: China Environmental Science Press.
Report on the State of the Environment in China, 1989–1990. Beijing: National Environmental Protection Agency.
Research Institute of Population Theory at Chinese People's University. 1987. *Population and Development*. Beijing: People's University of China Press.
"Resurgence of the Chinese Nation." Political Department of the PLA. *Chinese Population Newspaper*, March 16, 1990.
Robinson, H., et al. 1985. *Population and Resources*. Chen Jingtang, et al., trans. Beijing: Higher Education Press.
Shi Xilai. 1989. "Projections on Our Country's Population Trends in Light of the 1.90 Program." *Science and Technology Guide*, no. 4.
Shi Yulin, et al. 1985. *Wildlands in China Fit for Cultivation*. Beijing: Science and Technology Press.
Simon, Julian L. 1984. *The Economics of Population Growth*. Pen Songjing, et al., trans. Beijing: Beijing University Press.
———. 1985. *Growth Without Limit*. Huang Jiangan, et al., trans. Chengdu: Sichuan People's Press.
Song Jian. 1981. "A Quantitative Study of China's Ideal Population Objective and Its Optimization Analysis." A collection of papers submitted to the Third National Symposium on Population Science in Beijing, 1981, vol. 176, p. 183. Beijing: People's University of China Press.
———. 1982. *Population Projection and Population Control*. Beijing: Beijing People's Press.
———. 1985. *On Population Control*. Beijing: Science Press.
Stonepool, P. M. 1989. "Views on Chinese Agriculture." *Chinese Land Science*, no. 4.
Sun Jinxin. 1990. "Studies and Countermeasures on Controlling Population Growth." *Population Research*, no. 2.
Wang Xiaoshan. 1989. "Social Population and Natural Resources." *Population Front*, no. 2.
Wei Jinsheng. 1988. "Reflections on Several Basic Problems in the Control of China's Population Growth Over the Past Ten Years." *Population and the Economy*, no. 6.
"Wind Erosion and Desertification in Er Duosi Grasslands: Prevention and Control." 1988. *Journal of Natural Resources*, no. 3.
The World Commission on Environment and Development. 1989. *Our Common Future*. Beijing: World Knowledge Press.
"World Population at the Turn of the Century." 1989. Beijing: Research Center for Chinese Population Information.
Wu Cangping, et al. 1988. "Re-evaluation of the Characteristics of Population in the Context of Our National Conditions." *Population and the Economy*, no. 3.
Wu Mingyu, et al. 1989. *China and the World in the 1990s*. Beijing: China Finance and Economy Press.
Wu Zhongyun. *A History of Ideology on China's Population*. Beijing: China Social Sciences Press.
Xia Weisheng. 1984. *A Preliminary Exploration of Population Ecology*. Lanzhou: Gansu People's Press.
Xiao Chuan Yi Gao. 1991. "On Territorial Management in China." *Chinese Land Science*, no. 1.

Xiao Yuqi. 1989. "A Few Questions on the Environmental Capacity of Population." Paper presented at the First National Symposium on Population and Ecology in Nanjing, 1988.
Xu Songlin. 1988. "A Theoretical Analysis of Forest and Water Resources." *Journal of Natural Resources*, no. 4.
Xu Zhenglong, et al., eds. 1989. *A Green Report*. Beijing: Chinese Environment Newspaper Office.
Yan Hong. 1989. "Population Quality and Environmental Factors." Paper presented at the First National Symposium on Population and Ecology in Nanjing, 1988.
Yang Jingxing. 1989. "A Preliminary Exploration of a Theory on Population Ecology." Paper presented at the First National Symposium on Population and Ecology in Nanjing, 1988.
Yu Mouchang. 1989. "Problems in Population Ecology." Paper presented at the First National Symposium on Population and Ecology in Nanjing, 1988.
Yu Wentao, et al. 1987. *Environmental Protection in China*. Beijing: Science Press.
Yuan Winglin. *A History of Environmental Protection in China*. Beijing: China Environmental Science Press.
Zhang Changcai. 1990. "Breaking the Cycle: A Report from a Testing Area to Aid Environmental Projects in Bijie." *Chinese Population Newspaper*, March 2, 1990.
Zhang Jun. 1989. "An Analysis of the Constraints of Surplus Labor in Rural Areas." *Economics of China's Rural Areas*, no. 2.
Zhang Mincai. 1988. "Nature Sends a Warning to Man." *Population and the Economy*, no. 6.
Zhang Zhengning, et al., trans. 1989. *Economic Demography*. Beijing: Beijing Economy College Press.
Zhao Wenlin and Xie Shujun. *A History of China's Population*. Beijing: People's Press.
Zhao Xuan. 1989. "China Makes National Surveys of Birth and Birth Control." *Science and Technology Guide*, no. 4.
Zhao Xuedong. 1988. *The Challenge of Population Expansion and Its Countermeasures*. Chengdu: Sichuan University Press.
Zhen Jianmin. 1989. "Population Problems and Ecological Balance." *Population Information and Studies*, no. 3.
Zhou Kiaozheng. 1989. *Population Crisis*. Beijing: China Women's Press.
Zhou Lisan. 1991. "How to Best Use China's Resources." Paper presented at the Seminar for Resource Potential and Effective Measures for Efficient Use in Beijing, 1990.
Zhou Xiaozheng. 1989. "Improving Population Characteristics Is the Key to Our Country's Population Problems." *Science and Technology Guide*, no. 4.
Zhou Zhuliu. 1989. "Population Growth Must be Coordinated with Economic Development." *Population Front*, no. 3.
Zhu Baoshu. 1989. "The Carrying Capacity of the Environment." Paper presented at the First National Symposium on Population and Ecology in Nanjing, 1988.
Zhu Chuling, et al. 1988. "The Population Problem in Societies." Beijing: Science Awareness Press.
Zhu Guohong. 1989. "Modernization and China's Population." *Population Research*, no. 2.

Index

About the Book and Authors

Professors Qu and Li incorporate the results of historical research, current analysis, and forecasting to discuss the relationship between human population and the environment in China. Proposing ways that the PRC can move from vicious to positive cycles, they offer creative recommendations for overcoming the current crisis and promoting development.

A valuable scientific basis for China's policy decisions, the book is also an important reference for other nations, international organizations, and research institutes seeking to solve population and environmental problems.

Qu Geping is professor at Beijing, Tsinghua, Tongji, Wuhan, and Nanjing universities and at the People's University of China. He is also chair of the Environmental Protection Committee of the Chinese National Peoples Congress. He was previously administrator of China's National Environmental Protection Agency. The first Chinese representative to UNEP, Dr. Qu has held numerous distinguished positions, including secretary-general of the Chinese Environmental Science Society, vice-secretary-general and director of the Research Center for Policies on the Environment and Economy, and vice-director of China's Council for International Cooperation on the Environment and Development. Among his publications are *Chinese Environmental Problems, Strategies: China's Environmental Management*, and *China's Environment and Development*. He has won UNEP's Gold Prize and the UN Sasakawa International Environment Prize.

Li Jinchang is adviser to the Chinese National Environmental Protection Agency, first deputy director of the Contemporary Environmental and Economic Policy Research Center, and a member of the standing committee of China's Environmental Science Society. He is also professor at the Chinese Academy of Social Sciences, the People's University of China, and Chongqing University. He has served as China's deputy representative to UNEP. His publications include *China's Resources and Environment, Theory on Natural Resources Accounting*, and *Theory on Resource Industry*.